CHEMISTRY
TODAY

A crystal of the photographic fixer sodium thiosulfate, magnified about 40 times

The World Book
Encyclopedia of Science

CHEMISTRY TODAY

World Book, Inc.

a Scott Fetzer company

Chicago

ACKNOWLEDGMENTS

Consultant Editor Martin Sherwood

Consultants and Contributors
Harold Baum   John Bonner   Neil Carlson
Andrew Coghlan   Nigel Davis
Ronald Denney   C. A. Finch
Malcolm Frazer   Alan Katritzky
Percy Praill   Reginald Price
Martin Sherwood

Artists and Designers
Erick Drewery   Mick Gillah   Mick Saunders

Bull Publishing Consultants Ltd
Wendy Allen   Harold Bull   John Clark
Eric Drewery   Kate Duffy   Martyn Page
Polly Powell   Hal Robinson   Sandy Shepherd

Color Separation by
SKU Reproduktionen GmbH & Co. KG,
Munich

Setting
Typoservice Strothoff GmbH, Rietberg

CONTENTS

PREFACE

Chemistry Today, like the other volumes in this series of publications about the sciences, deals with a specific scientific area. This volume is concerned with chemistry. The subject is introduced with a discussion of its fundamental concepts: molecules, elements, formulas and key chemical reactions. Then follows an account of the principal groups of elements. Finally, each of the last three sections deals with a major speciality within the discipline of chemistry as a whole: organic chemistry, biochemistry, and analytical chemistry.

The editorial approach

The object of the series is to explain for an average family readership the many aspects of science that are not only fascinating in themselves but are also vitally important for an understanding of the world today. To achieve this the books have been made straightforward and concise, accurate in content, and are clearly and attractively presented. They are also a readily accessible source of scientific information.

The often forbidding appearance of traditional science publications has been completely avoided. Approximately equal proportions of illustrations and text make even the most unfamiliar subjects interesting and attractive. Even more important, all of the drawings have been created specially to complement the text, each explaining a topic that can be difficult to understand through the printed word alone.

The thorough application of these principles has created a publication that encapsulates its subject in a stimulating way, and that will prove to be an invaluable work of reference and education for many years to come.

The advance of science

One of the most exciting and challenging aspects of science is that its frontiers are constantly being revised and extended, and new developments are occurring all the time. Its advance depends largely on observation, experiment, dispute and debate, which generate theories that have to be tested and even then stand only until they are replaced by better concepts. For this reason it is difficult for any science publication to be completely comprehensive. It is possible, however, to provide a thorough foundation that ensures any such advances can be comprehended — and it is the purpose of each book in this series to create such a foundation, by providing all the basic knowledge in the particular area of science it describes.

How to use this book

This book can be used in two basic ways.

The first, and more conventional, way is to start at the beginning and to read through to the end, which gives a coherent and thorough picture of the subject and opens a resource of basic information that can be returned to for re-reading and reference.

The second allows the book to be used as a library of information presented subject by subject, which the reader can consult piece by piece as required.

All articles are prepared and presented so that the subject is equally accessible in either way. Topics are arranged in a logical sequence, outlined in the contents list. The index allows access to more specific points.

Within an article scientific terms are explained in the main text where an understanding of them is central to the understanding of the subject as a whole. Fact entries giving technical, mathematical or biographical details are included, where appropriate, at the end of the article to which they relate.

There is also an alphabetical glossary of terms at the end of the book, so that the reader's memory can be refreshed and so that the book can be used for quick reference whenever necessary.

All articles are relatively short, but none has been condensed artificially. Most articles occupy two pages, but some are four, or occasionally six, pages long.

The sample two-page article opposite shows the important elements of this editorial plan and illustrates the way in which this organization permits maximum flexibility of use.

(A) **Article title** gives the reader an immediate reference point.

(B) **Section title** shows the part of the book in which a particular article falls.

(C) **Main text** consists of approximately 850 words of narrative information set out in a logical manner, avoiding biographical and technical details that might tend to interrupt the story line and hamper the reader's progress.

(D) **Illustrations** include specially commissioned drawings and diagrams and carefully selected photographs, which expand, clarify and add to the main text.

(E) **Captions** explain the illustrations and make the connection between the textual and the visual elements of the article.

(F) **Annotation** of the drawings allows the reader to identify the various elements referred to in the captions.

(G) **Theme images,** where appropriate, are included in the top left-hand corner of the left-hand page, to emphasize a central element of information or to create a visual link between different but related articles. In articles about the major groups of elements, for example, the theme images relate the elements being discussed to their neighbors in the Periodic Table.

(H) **Fact entries** are added at the foot of the last page of certain articles to give additional information relating to the article but not essential to an understanding of the main text itself.

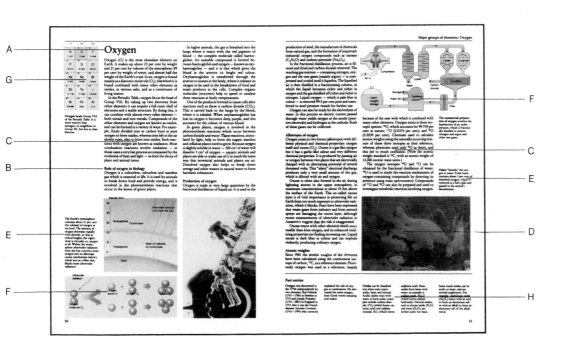

INTRODUCTION

For 4,000 years chemistry has been essential to the development of mankind. Since the beginning of the Bronze Age, people have used chemical processes in smelting and glassmaking; early medicine and alchemy were intimately related; and chemistry played a key role in the reawakening of scientific interest during the Renaissance and in the Industrial Revolution. But above all, it is in the last hundred years that the influence of chemistry has exploded, so that it now touches every aspect of scientific knowledge and civilized living.

Most of the familiar products around us depend on the chemical industry. Modern transportation relies on synthetic rubber, refined metals, and high-energy fuels. The construction industry needs paints, pigments, alloys, cements, glasses, plastics and ceramics. Our clothing and fabrics are increasingly manufactured from artificial fibers such as nylon and polyesters, colored by synthetic dyes, and cleansed by synthetic detergents and solvents. Fertilizers, antifreezes, disinfectants, pesticides, cosmetics, adhesives and drugs are just a few of the other products of synthetic chemistry.

The atomic theory of matter
The vital breakthrough that allowed chemistry to become the central science was the demonstration of the atomic nature of matter. In about 1800, John Dalton realized that all matter

Color photography relies on subtle light-induced reactions which are the province of photochemistry, a fast-growing specialized branch of the science of chemistry.

The white smoke pouring from the engine exhausts of aerobatic aircraft consists of finely divided titanium dioxide, which also finds uses in electronic components and in paint and paper making. The preparation and properties of metals and their compounds are part of the study of inorganic chemistry.

Gummy resin seeping from the seed pods of opium poppies contains morphine and related compounds. They belong to a group of plant-produced chemicals known as alkaloids and, like most other drugs and physiologically active substances, are dealt with in organic chemistry.

is composed of relatively few elements. He postulated that these elements combine together in accordance with their valency to make up discrete molecules. (Thus water, H_2O, is formed by the combination of two hydrogen atoms with one oxygen atom.) Antoine Lavoisier proved that burning is simply the chemical combination of the combustible material with the oxygen of the air. In the early nineteenth century, Friedrich Wöhler synthesized urea from inorganic materials, and demonstrated that there is no difference in principle between inanimate substances and those of living matter.

The atomic theory allows us to treat all the properties of matter in terms of the molecules that make it up. This has enabled the development of chemistry (and physics) to progress rapidly. An explanation in terms of molecular structure allows scientists to carry out rational experiments to test the explanations. Furthermore, correct explanations of significant and desired properties lead to the possibility of predicting improvements to those properties by molecular modification. This, in turn, then allows additional experiments to produce materials with such improved properties.

This chain of events is the basis of all applied research, and the chief reason why living standards have so improved that in the developed

nations the good life should now be possible for everyone.

Chemistry and the biological sciences

In the last 50 years, chemistry has become the language of the biological sciences and the basis of many of its most important experimental theories and methods. For example, reproduction is central to the concept of living matter. Since James Watson and Francis Crick broke the genetic code in the late 1950s, we understand in molecular terms how information is passed from one generation to another. Genes are made up of nucleic acids; nucleic acids allow the synthesis of proteins in living organisms; and proteins are the universal catalysts for the chemical processes that make up life.

The science of molecular genetics — literally genetics explained by chemistry — has already had great influence on world food production through the development of new improved strains of crop plants and farm animals. And much more progress lies just ahead.

Biochemistry — the chemistry of life

Biochemistry is the branch of organic chemistry that deals with the processes which take place in living matter. Physiology — plant, animal and human — is increasingly expressed in biochemical

terms. Today we know that blood carries oxygen around the body in chemical combination with the protein hemoglobin. We describe the working of nerves through chemical transmitters of known molecular formulae. Hormones are chemical messengers that monitor and control glandular functions. And the importance of the sodium-potassium balance in cell fluids is now well recognized.

Much of medicine is also dominated by chemical concepts. Medicinal chemists and pharmacologists design new pharmaceuticals for the control of diseases with reference to the molecular fitting of agent and protein receptor sites. In this way, the catalytic activity of the protein can be modified in the required direction. The recognition that sulfonamide drugs mimicked para-aminobenzoic acid derivatives led the way to the eradication of infections such as pneumonia. Present targets are the conquest of cancer and the understanding of the aging process.

The importance of inorganic chemistry

Inorganic chemistry is beginning to repay the great debt it owes to physics. Materials science is now based on molecular structure. The great advances in physics, electronics and engineering would be impossible without semiconductors, transistors and new mag-

The fabric of the balloons and the colors they are dyed with are just a few of the thousands of products of the chemical industry, which also provides metals, ceramics, plastics, fuels, medicines, cosmetics and even foodstuffs. Many are derived from petrochemicals, which are themselves made from petroleum in refineries such as the one illustrated on the opposite page.

netic materials — all crafted by chemical processes.

Astronomy relies increasingly on chemical concepts. A knowledge of the chemical composition of the planets, interstellar matter and the stars themselves is vital to test cosmological theories.

The geology of our planet Earth has been largely determined by chemical reactions in the formation of rocks. Much weathering is caused by chemical processes, particularly reactions with water and with carbon dioxide and oxygen from the atmosphere.

Chemistry and civilization

Modern civilization depends on the production of consumer goods, and many steps in manufacturing involve chemical modification. This applies to the production of mineral-based products in mining and refining, and no less to the production of animal- and plant-based products. Agriculture depends on chemically designed pesticides, fertilizers and growth promoters. Food production requires chemical analysis for quality control and the addition of preservatives, flavorings and coloring matter. Wood is increasingly modified before use as a constructional material. The production of paper is heavily dependent on chemical processing. And the purity of our water supplies depends on analysis and chlorination to kill harmful bacteria.

Why study chemistry?

The contribution of chemistry to other sciences has been emphasized in this Introduction. Most research in chemistry is now oriented towards the understanding and improvement of processes in some nonstrictly chemical area. However, fundamental research in chemistry is as exciting and important as ever. The chemists who painstakingly gathered butterflies to identify the pigment contained in their wings discovered the first compound of the type known as a pterin, a double-ring system containing four nitrogen atoms and six carbons. Some years later, pterins were found as essential components of all cellular matter — the earlier information was crucial in this recognition. In this way, much of the understanding depends on knowledge of fundamental chemical properties and structures.

Today, in every science, a knowledge of chemistry is important. The advances of the next hundred years will depend still more on relating the properties and functions of substances with their molecular composition.

Elements and molecules

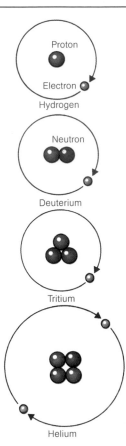

Simple models of simple atoms demonstrate the three main subatomic particles: the electron, proton and neutron.

Atomic orbitals are the regions in space where electrons are most likely to be found. An atom's first two electrons occupy an *s* orbital, the next six electrons are paired in three *p* orbitals, and the next 10 in five *d* orbitals. The orbitals have the shapes illustrated along the bottom of these pages.

Everything in the Universe — metals, stars, food, the air we breathe and we ourselves — is made up from combinations of only a few dozen different chemical elements. These, in turn, derive from only three different types of subatomic particles: protons, neutrons and electrons.

Each atom of an element consists of a dense nucleus surrounded by a diffuse cloud of electrons. Except in the case of the simplest element, hydrogen, the nucleus is always composed of protons and neutrons.

It is the number of protons in an atomic nucleus which establishes its identity. All atoms with just one nuclear proton are atoms of the element hydrogen; all those with two protons in the nucleus are atoms of helium, and so on. The number of protons is called the atomic number.

A proton has a positive electric charge and an electron has an equal negative charge. In a neutral atom, the numbers of protons and electrons are equal. The number of neutrons — which have no electric charge — may vary. This gives rise to different isotopes of the same element. For example, nearly all the hydrogen atoms in the Universe have only a proton in their nucleus, but a few also have one or two neutrons; these are the isotopes deuterium ("heavy hydrogen") with one neutron, and tritium with two.

In general the different isotopes of an element behave in a similar manner chemically, although their physical properties may vary slightly. Many of the elements with which we are familiar are made up from mixtures of several stable isotopes — the metal lead, for instance, is such a mixture. But some isotopes, such as those of uranium, are unstable; this gives rise to the phenomenon known as radioactivity.

Electron interactions

Chemistry is concerned with the way in which atoms interact. In particular, it is the study of the millions of different compounds that can be formed when atoms of different elements combine with each other into structures called molecules. Such interactions depend on the electrons rather than the nuclear particles of atoms.

Although the electron has been referred to as a particle, it frequently behaves like a wave. This makes it difficult to describe precisely how electrons are arranged around a nucleus. A simple analogy is to the orbits of planets. However, because of their wavelike properties, these orbits — or orbitals as they are called in chemistry — mark out a volume of space in which the electron may be found somewhere, rather than the two-dimensional ellipse marked out by a planetary orbit.

Nevertheless, electrons in different orbitals do have different average distances from the nucleus, just as planets are at different distances from the Sun. Because the attraction between the electrons and the nucleus is based on their opposite electric charges, the farther away an electron is from the nucleus, the less tightly is it held. The average distance of the electron from the nucleus is one of four different characteristics, called quantum numbers, that can be used to describe electrons in atoms.

In general, when atoms interact, they do so only through their outermost electrons, so it is these which are of most interest to chemists. Quantum numbers specify how many electrons can be at a particular energy level. In the case of hydrogen, for example, the level where its single electron occurs can hold a maximum of two electrons. If there are two electrons in this orbital, it is more difficult to remove an electron than if there is only one. This is why helium, with two protons and two electrons in its neutral state, is very unreactive. This also explains why hydrogen atoms will react with a great variety of other atoms.

Two atoms of hydrogen (H) can combine together to form a molecule of hydrogen (H_2) in such a way that each contributes its single electron to a molecular orbital. This orbital provides each nucleus with a share in both electrons, thus making the combination more stable than the two separate atoms. The molecular orbital occupies a volume of space which surrounds both nuclei and holds them together.

Molecular orbitals

The hydrogen molecule is called homoatomic, because it is made up from only one kind of atom. If a hydrogen atom joins with a fluorine atom (F),

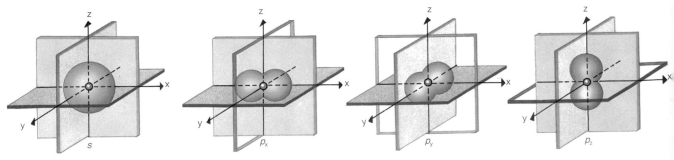

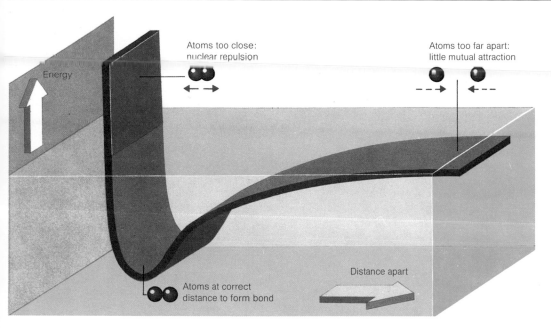

Atoms too close: nuclear repulsion

Atoms too far apart: little mutual attraction

Energy

Atoms at correct distance to form bond

Distance apart

A stable chemical bond between two atoms is formed when they are an optimum distance apart — a position of minimum energy. When they are too far apart, in a higher energy state, there is little mutual attraction and no bond formation. If they approach too close to each other, also in a high energy state, the positive charges on their nuclei repel each other and again no stable bond is formed. At the correct interatomic distance their atomic orbitals overlap to form a molecular orbital.

the resultant molecule of hydrogen fluoride (HF) is called heteroatomic.

In its outermost energy level, a fluorine atom has seven electrons out of a possible maximum of eight and it attracts electrons very strongly from other atoms. In the hydrogen fluoride molecule, the molecular orbital is not symmetrical, but pear-shaped, biased toward the fluorine nucleus. In other words, the pair of electrons that form the chemical bond spend more time closer to the fluorine nucleus than to the hydrogen nucleus.

Nevertheless, in both the hydrogen molecule and the hydrogen fluoride molecule, the electrons are shared. This type of chemical bond is called covalent.

If a highly electronegative atom, such as fluorine, meets a highly electropositive atom, such as sodium (Na), it may strip an electron completely away from the latter and form sodium fluoride (NaF). This molecule consists of a negatively charged fluoride ion (F^-) and a positively charged sodium ion (Na^+). The two ions are held together by electrostatic forces. This type of chemical bonding is called ionic.

There is a third way in which electrons may form a bond. It is a covalent bond in which both of the electrons originate from one of the atoms. This type of bonding is called coordinate bonding and is particularly important among the so-called transition elements, which include many metals.

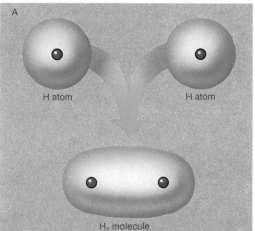

A

H atom

H atom

H_2 molecule

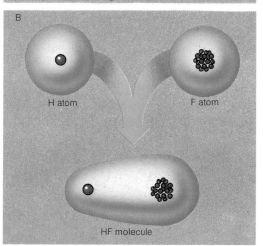

B

H atom

F atom

HF molecule

When two hydrogen atoms get close enough (A), their atomic orbitals overlap and form a molecular orbital. The electrons, one from each hydrogen atom, are shared equally in the resulting covalent bond in a hydrogen molecule. When hydrogen forms a similar bond with an atom of fluorine (B), the greater electronegativity of fluorine results in an asymmetric pear-shaped molecular orbital, with the fluorine atom having a larger share of the bonding electrons.

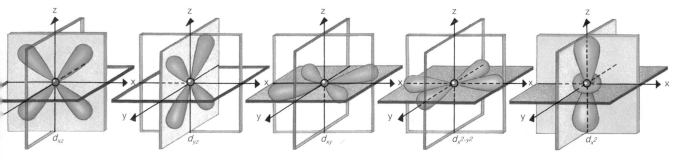

d_{xz} d_{yz} d_{xy} $d_{x^2-y^2}$ d_{z^2}

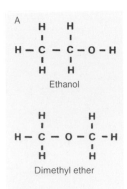

A

Ethanol

Dimethyl ether

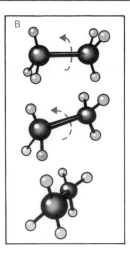

B

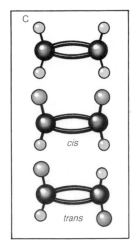

C

cis

trans

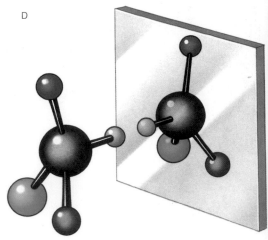

D

Isomers are pairs of molecules containing the same elements but in different spatial arrangements. In structural isomers (A) the atoms are joined in a different sequence, giving two totally different substances. In ethane, $CH_3—CH_3$, the two methyl (CH_3) groups can rotate about the carbon-carbon single bond, but interaction between the hydrogens gives it the shape shown (B). In ethene, $CH_2=CH_2$, the carbon-carbon double bond prevents such rotation (C). As a result, the related compound with, for example, two chlorine atoms replacing two hydrogens, exists as geometrical isomers known as *cis* (same side) and *trans* (opposite side) forms. A carbon atom bonded to four different atoms or groups (D) exists as two optical isomers which are mirror images of each other.

At its simplest, chemical combination can be likened to a child's construction kit, in which many different models can be made from only a handful of component pieces. In this diagram, as in others in the book, each colored sphere represents an atom and its accompanying cloud of electrons. The scheme gives various examples of elements and simple compounds, with their names and chemical formulas. (The colors are arbitrary and some of the molecules are three-dimensional, not flat.)

Chemical formulas

For every element there is an internationally agreed symbol which can be used as shorthand for its name. The simplest symbols are the first letter of the name. Thus the elements hydrogen, carbon and oxygen are abbreviated to H, C and O.

As there are many more elements than letters of the alphabet, most elements have to have two-letter symbols. Chlorine, the next heaviest element after carbon which begins with the letter C, is designated Cl. Calcium is Ca and cadmium, with the first two letters the same as in calcium, has the symbol Cd (see the table on page 15).

Most of the symbols for the elements are derived from their English names. However, in some cases, the symbols come from the Latin. Thus, sodium has the symbol Na *(natrium)* and lead has the symbol Pb *(plumbum).*

The composition of simple molecules can be shown by combining the symbols for the elements which they contain. Hydrogen fluoride, the molecule which contains one atom each of hydrogen and fluorine, can be written HF. In many molecules, more than one atom of the same

element is found — as with the hydrogen molecule. The number of atoms of any one element in a molecule is shown by a subscript number, as in H_2 for hydrogen.

Water molecules each contain two hydrogen atoms and one oxygen atom, represented as H_2O. This formula does not tell us how the atoms are joined together; but as hydrogen can form a chemical bond with only one other atom, whereas oxygen can form two, it is easy to work out that the water molecule has its oxygen atom in the middle, with a hydrogen atom at each end.

Structural isomers

Such representations are important in indicating the structure of complex molecules. Consider, for example, the formula C_2H_6O. Carbon atoms can form four covalent chemical bonds. As a result, this formula can be written in two different ways, in which all the atoms are joined by single bonds, to represent two quite different chemical compounds. One is ethanol, the alcohol found in drinks such as whiskey and brandy; the other is a close relative of the anesthetic ether. Compounds which contain the same number and types of atoms in different molecular arrangements are known as structural isomers.

Consequently, in complex compounds, it is important to give a clearer indication of structure than one obtains from a formula such as C_2H_6O. One way to do this is to show parts of the molecule separately. In ethanol, it is the —O—H part which gives the molecule its characteristic properties. Thus a more informative representation of the simple formula is C_2H_5OH.

The ether can be shown as $(CH_3)_2O$. Some groups of atoms occur so frequently in chemical compounds that they have been given their own special abbreviations. Thus, the combination of a carbon atom and three hydrogen atoms, known as the methyl group, is often written as Me. So this ether can also be shown as Me_2O.

Geometrical isomerism

Another complication in showing many chemical formulas is that atoms may be joined together by more than a single bond. This happens most frequently with carbon atoms, although a number of

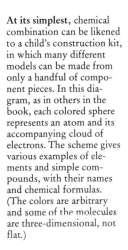

Elements	Compounds
○ Hydrogen	Water H_2O
● Carbon	Methane CH_4 Ammonia NH_3
● Nitrogen	Carbon dioxide CO_2
● Oxygen	Sodium hydroxide NaOH
○ Sodium	Hydrogen fluoride HF Magnesium chloride $MgCl_2$
● Potassium	Calcium sulfate $CaSO_4$ Potassium phosphate K_3PO_4
● Magnesium	
● Calcium	
○ Aluminum	
● Phosphorus	Magnesium nitrate $Mg(NO_3)_2$ Aluminum chloride $AlCl_3$
● Sulfur	
● Fluorine	
○ Chlorine	

To a diner, the salt (far left) is a condiment to put on his food, and to the road-builders (near left) sand has to be excavated to make a road-bed. Salt and sand are chemical compounds which a chemist calls sodium chloride and silica and gives the formulas NaCl and SiO_2. Thus each molecule of salt contains a sodium atom combined with a chlorine atom, and each silica molecule is made up of a silicon atom in chemical combination with two oxygen atoms.

other elements — oxygen and nitrogen, for example — are able to form multiple bonds.

The simple compound ethene (ethylene) consists of two carbon atoms held together by a double bond, and four hydrogen atoms. Two of these hydrogens are singly bonded to one carbon atom, the other two to the other carbon. The carbon-carbon double bond can be represented in two ways, as —C=C— or as —C:C—. Thus ethene is written as $H_2C=CH_2$ or $H_2C:CH_2$, which conveys more information about the structure of the molecule than does the simple formula C_2H_4.

Once multiple bonds are taken into account, a new problem emerges. Molecules exist in three dimensions, not two. They have shapes that can change only within certain limits. If we look at another simple two-carbon molecule, ethane, we find this is made of two carbon atoms, joined by a single bond, and six hydrogen atoms (three attached to each carbon atom). It can be written as $H_3C—CH_3$ (or $H_3C.CH_3$). Diagram "B" (opposite) shows a way of representing this molecule which indicates the effect of its electron clouds. These clouds, being negatively charged, tend to repel one another. Consequently the ethane molecule has the hydrogen atoms on one carbon out of alignment with those on the other; this keeps the negative charges as far apart as possible.

Nevertheless, the carbon atoms can rotate about their single bond. Only a small amount of energy is needed to overcome the reluctance of the electron clouds to come slightly closer together. A second bond linking two carbon atoms prevents such rotation, and this means that molecules with the same basic structure may occur in chemically distinct forms, called isomers. This type of isomerism, geometric isomerism, is important in many living processes.

Optical isomerism

Another important type of isomerism is optical isomerism, when two molecules of a substance are mirror images of each other. They are so called because they are optically active — that is, their solutions have the property of rotating the plane of polarized light. Such isomers occur most frequently in carbon chemistry, in compounds in which a carbon atom is tetrahedrally bonded to four different atoms or groups. The two forms are generally very difficult to separate; many of their chemical properties are identical. But often only one of the isomers is biologically active.

The table below lists alphabetically the chemical elements and gives their symbols, which are used in chemical formulas and in the Periodic Tables in the next section of this book.

Actinium	Ac	Cobalt	Co	Iron	Fe	Palladium	Pd	Strontium	Sr
Aluminum	Al	Copper	Cu	Krypton	Kr	Phosphorus	P	Sulfur	S
Americium	Am	Curium	Cm	Lanthanum	La	Platinum	Pt	Tantalum	Ta
Antimony	Sb	Dysprosium	Dy	Lawrencium	Lr	Plutonium	Pu	Technetium	Tc
Argon	Ar	Einsteinium	Es	Lead	Pb	Polonium	Po	Tellurium	Te
Arsenic	As	Erbium	Er	Lithium	Li	Potassium	K	Terbium	Tb
Astatine	At	Europium	Eu	Lutetium	Lu	Praseodymium	Pr	Thallium	Tl
Barium	Ba	Fermium	Fm	Magnesium	Mg	Promethium	Pm	Thorium	Th
Berkelium	Bk	Fluorine	F	Manganese	Mn	Protactinium	Pa	Thulium	Tm
Beryllium	Be	Francium	Fr	Mendelevium	Md	Radium	Ra	Tin	Sn
Bismuth	Bi	Gadolinium	Gd	Mercury	Hg	Radon	Rn	Titanium	Ti
Boron	B	Gallium	Ga	Molybdenum	Mo	Rhenium	Re	Tungsten	W
Bromine	Br	Germanium	Ge	Neodymium	Nd	Rhodium	Rh	Uranium	U
Cadmium	Cd	Gold	Au	Neon	Ne	Rubidium	Rb	Vanadium	V
Calcium	Ca	Hafnium	Hf	Neptunium	Np	Ruthenium	Ru	Xenon	Xe
Californium	Cf	Helium	He	Nickel	Ni	Samarium	Sm	Ytterbium	Yb
Carbon	C	Holmium	Ho	Niobium	Nb	Scandium	Sc	Yttrium	Y
Cerium	Ce	Hydrogen	H	Nitrogen	N	Selenium	Se	Zinc	Zn
Cesium	Cs	Indium	In	Nobelium	No	Silicon	Si	Zirconium	Zr
Chlorine	Cl	Iodine	I	Osmium	Os	Silver	Ag		
Chromium	Cr	Iridium	Ir	Oxygen	O	Sodium	Na		

There are four main types of chemical reactions, which can be generalized by the simple equations above. In (1) a single substance A undergoes a chemical change to produce a different substance B. In (2) the single substance A decomposes to form two new substances B and C. In reaction type (3), two substances A and B combine to form a third, new substance C. And finally in (4), the reactants A and B take part in a chemical reaction to form the two products C and D. Reactions of type (4) are the most common, and may involve more than two reactants or give more than two products.

Key chemical reactions

The world around us changes continually. Some of this change is purely physical. When water boils, for example, the liquid water turns to gaseous steam; in the sub-zero temperatures of winter, a puddle turns to solid ice. These are both purely physical changes. Altering the temperature alters the physical form of the water, but the constituent molecules themselves remain unchanged: each is still composed of two hydrogen atoms linked to an oxygen atom (H_2O).

On the other hand, if we pass an electric current through water, a mixture of gases is produced. Under everyday conditions, we cannot turn these gases into liquids, let alone solids. The molecules that make up these gases are no longer water molecules; they are a mixture of hydrogen (H_2) and oxygen (O_2). If a spark is applied to the mixture, it explodes as the gases recombine to produce water. Both of these are chemical, not physical, changes.

This behavior demonstrates two important principles of chemistry. Firstly, under some conditions, substances change into wholly different substances. A chemical change — or reaction — is one in which the molecules are not the same at the end of the process as they were at the beginning. Secondly, if we can change water into different substances, then change these back to water, at least one chemical reaction must be reversible. In fact, under the right conditions, nearly all chemical reactions are reversible.

Molecules and energy
When a current passes through water, the energy from the electricity breaks the water molecules apart into hydrogen and oxygen atoms. The atoms of each type then combine to form diatomic molecules of hydrogen and oxygen; both substances are gases at ordinary temperatures. A spark applied to a mixture of the two gases triggers off a reaction between them which makes them reform water molecules and release energy (in the form of light, heat and sound).

In chemical terms, a reaction occurs spontaneously only if the products are less energetic than the reactants — that is, if the products are chemically more stable. Several factors contribute to the energy of a molecule, so a spontaneous reaction does not always liberate energy in a recognizable form, such as light and heat. Nevertheless, many reactions do give off large quantities of usable energy. When natural gas burns, for example, its molecules react with oxygen, releasing energy which we can use to heat our homes and cook our food.

Anyone who has seen gas burning realizes that energy is released in the process. But can we call it spontaneous? The gas does not combust by itself. Only when it is mixed with air and a spark or match applied does the reaction occur. Why is this?

When a reaction takes place, chemical bonds are broken and new ones formed. In the case of the electrolysis of water, the bonds between the hydrogen and oxygen atoms in water break and new bonds form between pairs of hydrogen atoms and pairs of oxygen atoms. In the case of natural gas, which consists mainly of the hydrocarbon methane (CH_4), it is bonds between carbon and hydrogen which break, and new bonds between oxygen and carbon and oxygen and hydrogen which form.

If we use chemical formulas to show such reactions, they appear like this:
$$2H_2O \rightarrow 2H_2 + O_2$$
$$CH_4 + 2O_2 \rightarrow CO_2 + 2H_2O$$
The equations are similar to those used in mathematics, in that both sides of an equation must balance. That is, they contain the same number and types of atoms, but differently connected.

Although such equations are useful, they oversimplify the nature of a chemical reaction. What generally happens during a reaction is that an intermediate or transition state forms. Intermediates are usually very unstable and cannot be isolated; they break down rapidly, either back to the starting materials or into the products of the reaction. They are so unstable because they are more energetic than either the reactants or products. To form them, it is usually necessary to supply energy to a reaction.

If the reaction releases energy, then provided that a few molecules get sufficient energy to reach the transition state, they release enough energy when they break down into products to form more of the transitional species and keep the process going. If the reaction releases a lot of energy, then a very small trigger — such as the spark in a gasoline engine — can be enough to cause an explosively fast reaction.

On the other hand, if you apply a spark to a piece of paper, it is not a sufficient energy input to make the reaction self-sustaining. But apply a

Electrolysis is a key chemical reaction. When a direct current (DC) is passed through water (containing a little acid to make it a better conductor of electricity), oxygen gas is evolved at the positive electrode (anode) and hydrogen is evolved at the cathode. The volume of hydrogen is exactly twice the volume of oxygen, in accordance with the chemical formula for water, H_2O, and the equation
$$2H_2O \rightarrow 2H_2 + O_2$$

Acidified water

Hydrogen

Oxygen

Platinum cathode

Platinum anode

Many of the chemical reactions important to today's industry originally took place millions of years ago. In the potash mine (left), more than 3,000 feet (915 meters) beneath the North Sea, a mechanical digger grinds out ancient sedimentary deposits of potassium chloride, an essential ingredient of artificial fertilizers.

In the commercial process for producing chlorine (below), molten sodium chloride is electrolyzed using a mercury cathode, which carries off the sodium that is also produced. Reacting the mercury/sodium amalgam with water gives sodium hydroxide.

lighted match and, once the paper has caught fire, it continues to burn steadily. The molecules in paper, as they react with oxygen, release enough energy to make a few more molecules react, but not enough to make the reaction go explosively fast.

The idea of the transition state between reactants and products helps to explain why reactions are reversible. If enough energy is supplied to the starting materials (reactants) of a reaction, they are able to reach the transition state. It is always possible for an intermediate species to break down into either reactants or products. If energy is being dissipated during a reaction — heat being given off, for example — then lower-energy products find it difficult to reform intermediate species. If energy is being constantly supplied, the reverse is true. That is why, when a piece of iron is left in the open, it gradually turns to rust — iron

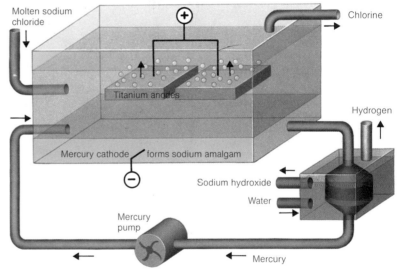

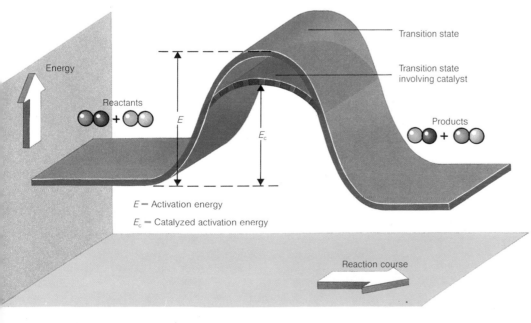

E = Activation energy

E_c = Catalyzed activation energy

Reaction course

Initiating a chemical reaction generally requires the injection of enough energy for some of the reactants to form an intermediate stage called a transition state. The reaction then proceeds, with the overall energy content of the products being less than that of the reactants. A catalyst can have the effect of lowering the energy barrier, so making the reaction proceed more readily.

Michael Faraday, shown here working in his laboratory, was one of the first great experimental chemists. He formulated the laws of electrolysis, discovered benzene and other organic substances, and first demonstrated the use of platinum as a catalyst.

oxide — but when large quantities of iron oxide are heated with other materials in a smelting furnace the product is metallic iron.

Oxidation-reduction reactions

The few simple reactions considered so far all involve oxidation or its converse, reduction. Such reactions are among the most important in chemistry. Many processes that take place in living organisms, for example, depend on a whole series of interdependent oxidation-reduction reactions.

Despite its name, an oxidation reaction does not necessarily involve oxygen atoms. The chemical definition of an oxidation reaction is one in which the "oxidation state" or "oxidation number" of an atom increases; a reduction is one in which the oxidation state decreases.

Although most compounds are made up of atoms joined together by covalent bonds, the oxidation state is calculated by assuming that they

can be treated as ions — atoms carrying positive or negative electrical charges. In its neutral state, an atom is always considered to have an oxidation state equal to zero.

If we consider oxygen in water, it has shares in two additional electrons. If water were an ionic compound, we could imagine the oxygen having a double negative charge. But because the molecule is electrically neutral overall, the hydrogen atoms have to be assumed to be lacking an electron and therefore have a single positive charge. (So far as oxidation number is concerned, it is always assumed that the more electronegative element in a combination is assigned the negative charge.)

On this basis, when water forms from hydrogen and oxygen gases, the hydrogen atoms move from an oxidation state of zero to one of $+1$: they are said to have been oxidized. Conversely, the oxidation state of the oxygen atoms goes from zero to -2; they have been reduced. The formation of water thus involves simultaneously a reduction and an oxidation. This is true also of many more complex reactions.

Multiple oxidation states

Let us look at some molecules in isolation. Nitrogen can form compounds with oxygen as well as hydrogen. Ammonia consists of molecules in which a single nitrogen atom is attached to three hydrogen atoms (NH_3). Nitrogen is a more electronegative element than hydrogen, and so it is said to have an oxidation number of -3 in ammonia. In dinitrogen pentoxide (N_2O_5), however, where oxygen is the more electronegative element, the nitrogen's oxidation number is $+5$. Where an element can exist in many different oxidation states, it increases the number of possible combinations it can form with other atoms. Thus nitrogen can form a whole range of compounds with oxygen and hydrogen, in which its oxidation state takes on all possible values between $+5$ (in dinitrogen pentoxide) and -3 (in ammonia), including 0 (in free nitrogen gas).

Dinitrogen pentoxide	N_2O_5		$+5$
Nitrogen dioxide	NO_2		$+4$
Nitrogen tetroxide	N_2O_4		$+4$
Dinitrogen trioxide	N_2O_3		$+3$
Nitric oxide	NO		$+2$
Nitrous oxide	N_2O		$+1$
Nitrogen	N_2		0
Hydroxylamine	NH_2OH		-1
Hydrazine	N_2H_4		-2
Ammonia	NH_3		-3

Nitrogen, with its oxides and hydrides, demonstrates a wide range of oxidation states from $+5$ (in dinitrogen pentoxide, N_2O_5) to -3 in ammonia (NH_3). This diagram also illustrates the shapes of the various molecules (brown = nitrogen atom; red = oxygen atom; yellow = hydrogen atom).

To demonstrate the point made earlier that an oxidation reaction need not involve oxygen, consider the behavior of molten sodium chloride (NaCl, common salt) when an electric current is passed through it. Sodium chloride is made up from atoms of a strongly electropositive element and a strongly electronegative element. As a result, it is an ionic compound: chlorine exists as ions with a single negative charge (oxidation number -1) and sodium as ions with a single positive charge (oxidation number $+1$). At the negative electrode, electrons from the current are attracted to the sodium ions (Na^+) and change them into neutral sodium atoms (a reduction reaction, because the oxidation number decreases). The chlorine ions (Cl^-) are, however, attracted to the positive electrode in the circuit, where their extra electrons are stripped off, thus oxidizing them to neutral chlorine atoms.

For elements and some stable groupings, it has been possible to work out standard oxidation-reduction potentials. These are the voltages needed to convert, say, a neutral metal atom to its positive ion. The importance of this series is that, by selecting ions of one element and mixing them with neutral atoms of another, an oxidation-reduction takes place and generates a voltage. This is the basis of an electric cell or battery. For example, if metallic copper is placed in a solution of zinc ions, a cell is formed which generates about 1 volt, while accompanied by an electrochemical reaction in which zinc is reduced to the neutral metal and copper is oxidized to copper ions.

Other reactions

If we express the conversion of copper to copper sulfate in a solution of zinc sulfate as an equation, we get

$$Cu + ZnSO_4 \rightarrow CuSO_4 + Zn$$

Combustion takes many forms, all of which are examples of oxidation. Burning (above) is, in chemical terms, a comparatively slow reaction although often extremely destructive. Even the explosively fast reactions involved in firing a pistol (below) take place in stages. The percussion cap at the base of a cartridge (A) is detonated by a hammer or firing pin (B) and ignites the propellant charge (C), which "burns" to produce a large volume of hot gas which drives the bullet out of the barrel (D).

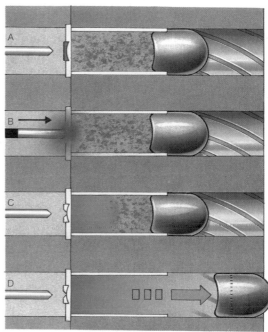

Smelting of metals (above) involves chemical reduction and often requires large amounts of energy to "drive" the reactions — for example the conversion of iron oxide to metallic iron. The corresponding oxidation reaction, as in the rusting of iron (above right), is energetically favorable and proceeds spontaneously in moist air.

The sulfate ion (SO_4^{2-}) is a stable complex ion which remains unchanged throughout the reaction. In effect, it shifts from one metal atom to another, causing an oxidation and a reduction at the same time. However, most chemical reactions take place between compounds rather than between elements or elements and compounds. In such cases, it is often possible for atoms to exchange the other atoms or groups of atoms to which they are attached. For example, if sodium sulfate (Na_2SO_4) is added to a solution of barium chloride ($BaCl_2$), the following equation demonstrates the reaction:

$$Na_2SO_4 + BaCl_2 \rightarrow BaSO_4 + 2NaCl$$

The barium sulfate formed is very insoluble and comes out of solution as a white precipitate, leaving behind a solution of sodium chloride (common salt).

Such "double decomposition" reactions, as they are called, depend on the ionic character of the compounds in solution. The ions of each compound split apart under the solvating influence of water. Where different ions can combine more strongly — as in the case of the barium and sulfate ions — and leave the solution, they ensure that the reaction goes to completion — that is, that the conversion of one compound into another is effectively complete.

A common type of double decomposition occurs when an acid is mixed with a base. In general, acidic compounds are those which, in solution, dissociate to form hydrogen ions (H^+), whereas bases (or alkalis, as the common ones are also called) form hydroxyl ions (OH^-). In fact, the acidity of a solution is measured in terms of the concentration of hydrogen ions. For example, if hydrochloric acid and sodium hydroxide are mixed, the result is a solution of common salt:

$$HCl + NaOH \rightarrow NaCl + H_2O$$

In acid-base neutralizations, the products are always a salt (the common name for an ionic compound) and water. The reaction occurs because molecules of water dissociate less easily than those of acids and bases and thus remove ions from solution in the same way that precipitation of barium sulfate does.

The elimination of the components of water from one or more reactants is very common. It occurs with many of the complex carbon-con-

Fact entries

Chemical reaction
For a reaction to occur between two different molecules, they must come very close together. If they have enough energy to reach a transition state, they may then react. When molecules are heated, they move around more freely; if they are in an enclosed space, this increases their chance of collisions with other molecules and thus the likelihood of a reaction occurring.

Chemical equilibrium
Under any set of conditions, there is a fixed chance that an intermediate species will break down one way rather than another. This leads to an equilibrium. Given enough time, the same percentage of starting materials is converted into products. However, equilibrium can be changed by altering the reaction conditions. The formation of the ester ethyl ethanoate from ethanoic

acid (acetic acid) and ethanol (ethyl alcohol) also produces water. If the reaction is carried out in such a way that water is removed as it forms, then the back reaction cannot occur. This can be done by adding concentrated sulfuric acid and heating the reactants.
Catalysis Many substances are absorbed strongly on surfaces; this phenomenon can be used to help reactions along. Metals, in particular,

are often used to catalyze reactions. Molecules concentrate on the metal surface, thus bringing them into close proximity and increasing the chances of reaching a transition state.

Catalysis is very important, both in living and industrial processes. A catalyst speeds up a reaction, but is unchanged by it. Surface catalysis by metals is only one type. In other types the catalyst

may react chemically with the molecule(s) involved in a reaction, but then be regenerated when the products form. Enzymes, the catalysts in living systems, frequently work in this way. They temporarily bind other molecules and hold them together so that they are in just the right spatial relationship to react together to form products. The enzyme, in effect, lowers the activation energy barrier.

taining molecules found in organic chemistry. The formation of the synthetic fiber nylon, for example, occurs when many hundreds of molecules of one or two simple compounds react together to form long chains, eliminating a molecule of water as each link in the chain is forged. Other elimination reactions, in which different small molecules — for example ammonia (NH_3) — are removed also occur often. Where the reaction involves the remaining parts of the reactants joining together, it is also effectively an addition reaction. Conventionally, addition reactions do not involve removal of any part of the reacting molecule. This type of recombination is particularly common in the case of reactions involving carbon-carbon double or triple bonds.

The reverse of addition is decomposition, in which a single reactant breaks down into more than one product. A simple example of decomposition is calcium carbonate ($CaCO_3$, chalk). If this compound is heated strongly, it breaks down into a mixture of carbon dioxide gas (CO_2) and calcium oxide (CaO, quicklime). The reaction requires heating continuously to keep it going, and so it is termed endothermic. A more complicated decomposition reaction occurs on heating the substance ammonium dichromate, $(NH_4)_2Cr_2O_7$. As revealed by the equation for the reaction

$$(NH_4)_2Cr_2O_7 \rightarrow Cr_2O_3 + 2N_2 + 4H_2O$$

the compound decomposes into an oxide of chromium, nitrogen gas and water (evolved as steam). Unlike the calcium oxide decomposition, this is an exothermic reaction; once the reaction starts, the heat it produces sustains it without the need for continued external heating.

In some cases, a molecule may break down only partly and then reform chemical bonds in a different manner. Such reactions are called rearrangements and are often significant in synthetic chemistry and in the chemistry of living systems. An important part of the chemistry of vision, for example, depends upon such a molecular rearrangement — in this case, the breaking of a carbon-carbon double bond and its reformation to give a different geometrical isomer of the starting compound.

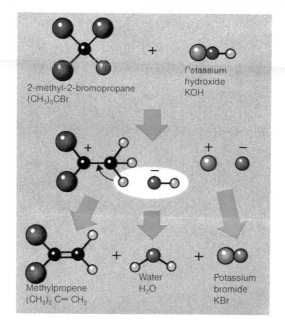

2-methyl-2-bromopropane $(CH_3)_3CBr$

Potassium hydroxide KOH

Methylpropene $(CH_3)_2 C = CH_2$

Water H_2O

Potassium bromide KBr

Elimination reactions are important in synthetic chemistry and involve the formation of a small simple molecule such as water or ammonia. The elimination of water (left) accompanies the conversion of an organic bromide into a double-bonded hydrocarbon (an alkene) in alkaline solution. In the manufacture of nylon (below) the elimination of water accompanies the polymerization reaction that creates the synthetic fiber.

Photochemical reactions
The energy to initiate a reaction comes, in many cases, from heat. The rate of most reactions increases regularly with temperature, roughly doubling for each 18°F (10°C) rise in temperature. However, other forms of energy can also initiate reactions. Photochemistry is the branch of chemistry concerned with the interaction between light (photons) and molecules. Such reactions are essential to life on Earth: the trapping of solar energy in chemical form by plants is the first step in the food chain for many living species.

Photochemical reactions are also the basis of photography. In black-and-white photography, the key reaction is the conversion of silver salts to finely-divided particles of silver (which appear black) when light strikes them. Color photography depends upon a range of complex light-sensitive carbon compounds.

In some cases, the effect of a very small amount of light can be dramatic. A mixture of hydrogen and chlorine gases kept in the dark at room temperature does not react. Exposure to light starts an explosive reaction, however. This is because a chlorine molecule (Cl_2) breaks into two chlorine atoms when it absorbs light energy. An unpaired chlorine atom is highly reactive and breaks up a hydrogen molecule to form a molecule of hydrogen chloride (HCl) and a reactive, unpaired hydrogen atom. The latter can break up a molecule of chlorine, producing another molecule of hydrogen chloride and a further chlorine atom. Thus, each step produces the reactive atom necessary for the next one. Such reactions are called chain reactions.

Major groups of elements

From a knowledge of the subatomic structures of the elements, it is possible to arrange them into groups which show similar behavior. This forms the basis of modern chemistry. As scientists gathered factual information about different elements during the eighteenth and nineteenth centuries, similarities began to appear between some of them. But not until the latter half of the nineteenth century was sufficient data available to make possible a serious attempt at classifying all the elements.

Even then, the classification was empirical; in other words, it was based on observation, there being no knowledge of subatomic particles at that time. The most successful proponent of such classifications was the Russian chemist Dmitri Mendeleyev, whose Periodic Table not only included all the elements known in his time, but also left gaps for ones which he predicted ought to exist. It was the discovery of some of these "missing" elements which led scientists to accept his basic scheme.

Regularities

With a modern knowledge of the electronic structure of the elements, it is easy to see why they fall into regular grouping (called periods). As the atomic number increases, so does the number of electrons surrounding the nucleus. At particular intervals, specified by quantum theory, a shell of electrons is completed. This confers stability on the structure, as we have already seen from the formation of molecules. Further, the extra electron of the next element starts a new shell. It has a single electron in this shell, which is relatively easy to remove.

Sodium and potassium are two elements that have a single electron in their outer shells. Both are metals that react violently with water. Both readily form ionic compounds, such as chlorides. Nevertheless, they are different elements. Potassium has one more complete shell of electrons between its outer electron and its nucleus. This makes its properties different in detail from those of sodium. The transmission of nerve impulses in living organisms, for example, depends on differences between sodium and potassium ions, not on their similarities.

If atoms are arranged into a table so that those with the same numbers of outer electrons are listed under each other, they form families with common characteristics. If we consider the elements with one more proton and electron than sodium and potassium (namely magnesium and calcium), we find that both form ionic compounds by the loss of two electrons — not surprisingly, because that is the total number of electrons in the outer shell.

Quantum numbers

The electrons in each element are characterized by four different quantum numbers, the first of which describes the average distance of an electron from the nucleus. This quantum number also controls the second and third quantum numbers, which define the shapes and orientations of the atomic orbitals (the regions in space occupied by the electrons). Different orbitals have different energies, but the difference in energy between electrons with the same first quantum number is generally less than that between electrons with different first quantum numbers.

The electron configuration of an atom is often referred to in a shorthand, which shows the number of electrons in each orbital. The first quantum number is indicated by a number. The second is denoted by a letter (s, p, d or f). The number of electrons in a particular orbital is shown as a

Few solid elements occur free in nature. Exceptions are unreactive metals such as gold (left) and silver, and the nonmetals carbon and sulfur (right). Nearly all other elements — apart from the gases in the atmosphere — occur in combination as compounds.

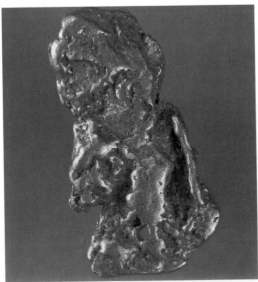

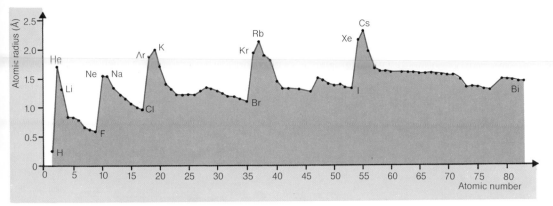

superscript. (The third and fourth quantum numbers are ignored in this notation.)

What the quantum rules say is that, when the first quantum equals 1, only an s orbital is allowed. When this quantum number equals 2, both s and p orbitals are allowed and, when it is 3, there can be s, p and d orbitals. The quantum rules also state the maximum numbers of electrons that can occupy the orbitals (two electrons for an s orbital, six for p and ten for d).

If we add up the allowed numbers of electrons in elements as the atomic mass increases, we find that argon has 18 electrons. Its configuration can be shown as $1s^2\ 2s^2\ 2p^6\ 3s^2\ 3p^6$. If the quantum rules were rigid, the element following argon would have one electron in its $3d$ orbital. But this is not the case. The 19th element is potassium, which has a single s electron in the fourth shell. By the time a nucleus is surrounded by this many electrons, the energy levels are sufficiently perturbed that the $4s$ orbital is of lower energy than

the $3d$ orbital. However, the $4p$ orbital is of higher energy than the $3d$ orbital. This leads to what is called the first "long period" in the Periodic Table, the one which involves filling of the $3d$ orbital while retaining two electrons in the $4s$ orbital.

Even this rule is not firm and variations from it affect the properties of the individual elements. Nevertheless, it is sufficient to provide most of the categorization on which modern chemistry is built.

It is this categorization which is used in the following articles to divide the elements into small family groups. Each article has an illustration, at the top left-hand corner, of the section of the Periodic Table that includes the element or elements being described. For the names of the elements whose symbols appear below, consult the table in the article on **Chemical formulas**, which appears in the first part of this book (see page 15).

Periodic Table of the elements

Arranging the elements in horizontal bands in order of increasing atomic number creates the Periodic Table. Elements with similar chemical properties form vertical groups. Each "box" has the element's chemical symbol at the center, with the atomic number top left and the atomic mass (mass number) below.

IA	IIA											IIIA	IVA	VA	VIA	VIIA	0
1 H 1.008																	2 He 4.003
3 Li 6.941	4 Be 9.012											5 B 10.810	6 C 12.011	7 N 14.007	8 O 15.999	9 F 18.998	10 Ne 20.179
11 Na 22.990	12 Mg 24.305	IIIB	IVB	VB	VIB	VIIB	VIII			IB	IIB	13 Al 26.982	14 Si 28.086	15 P 30.974	16 S 32.060	17 Cl 35.453	18 Ar 39.948
19 K 39.102	20 Ca 40.080	21 Sc 44.956	22 Ti 47.900	23 V 50.941	24 Cr 51.996	25 Mn 54.938	26 Fe 55.847	27 Co 58.933	28 Ni 58.700	29 Cu 63.546	30 Zn 65.380	31 Ga 69.720	32 Ge 72.590	33 As 74.922	34 Se 78.960	35 Br 79.904	36 Kr 83.800
37 Rb 85.468	38 Sr 87.620	39 Y 88.906	40 Zr 91.220	41 Nb 92.906	42 Mo 95.940	43 Tc (97)	44 Ru 101.070	45 Rh 102.906	46 Pd 106.400	47 Ag 107.868	48 Cd 112.400	49 In 114.820	50 Sn 118.690	51 Sb 121.750	52 Te 127.600	53 I 126.904	54 Xe 131.300
55 Cs 132.906	56 Ba 137.340	57 La* 138.906	72 Hf 178.490	73 Ta 180.948	74 W 183.850	75 Re 186.207	76 Os 190.200	77 Ir 192.220	78 Pt 195.090	79 Au 196.967	80 Hg 200.590	81 Tl 204.370	82 Pb 207.200	83 Bi 208.980	84 Po (209)	85 At (210)	86 Rn (222)
87 Fr (223)	88 Ra (226)	89 Ac** (227)	104 (261)	105 (262)													

58 Ce 140.120	59 Pr 140.908	60 Nd 144.240	61 Pm (145)	62 Sm 150.400	63 Eu 151.960	64 Gd 157.250	65 Tb 158.925	66 Dy 162.500	67 Ho 164.930	68 Er 167.260	69 Tm 168.934	70 Yb 173.040	71 Lu 174.970
90 Th 232.038	91 Pa 231.036	92 U 238.029	93 Np 237.048	94 Pu (244)	95 Am (243)	96 Cm (247)	97 Bk (247)	98 Cf (251)	99 Es (254)	100 Fm (257)	101 Md (258)	102 No (255)	103 Lr (260)

* Lanthanides

** Actinides

3 Li 6.941	4 Be 9.012
11 Na 22.990	12 Mg 24.305
19 K 39.102	20 Ca 40.080
37 Rb 85.468	38 Sr 87.620
55 Cs 132.906	56 Ba 137.340
87 Fr (223)	88 Ra (226)

Hydrogen, like the alkali metals (Group IA) immediately below it in the Periodic Table, has one electron in its outer shell. Unlike the alkali metals, however, hydrogen has only one electron shell.

Hydrogen

Hydrogen, symbol H, is the first element of the Periodic Table, and has the simplest atomic structure. It is the most abundant element in the Universe, although only the ninth most abundant on Earth. The hydrogen atom consists of a nucleus of one positive charge, with an outer shell of one electron. There are three isotopes: mass approximately 1 (sometimes known as "protium"), which comprises about 99.98 per cent of the natural element; deuterium, mass approximately 2 (symbol D or ^2H), in which the nucleus contains one proton and one neutron, is about 0.0156 per cent; and tritium, mass approximately 3 (T or ^3H) with one proton and two neutrons, is about 10^{-15} per cent. Tritium, believed to arise from slow radioactive disintegrations in rocks and minerals, can be produced artificially by atomic bombardment. It is radioactive, with a half-life of 12.26 years.

Hydrogen can be produced in quantity from water or renewable sources of energy and may, in the future, be used as an energy carrier to reduce dependence on fossil fuels.

Properties of hydrogen

As a gas, hydrogen is transported in steel cylinders at 120—150 atmospheres pressure, or as a liquid (boiling point about —423°F, or —253°C) in thermally insulated containers. Liquid hydrogen is used as a rocket fuel. The gas was once used for filling balloons and airships because, as the lightest gas, it has better "lift" than any other. A volume of 13,080 cubic yards (10,000 cubic meters) of hydrogen at 32°F (0°C) and 1 atmosphere pressure can lift about 13.5 short tons (12.2 metric tons). However it is inflammable, and airships are now filled with helium, the next lightest element.

An unusual property of the gas, also showing its lightness, is the increase in pitch of sound when hydrogen replaces the surrounding air. The velocity and therefore frequency of the sound varies with the density of the surrounding gas — so the pitch of the sound produced is higher than in air.

Compounds and uses of hydrogen

Hydrogen forms many compounds: it is a component of water, all acids and bases, most organic compounds, and many minerals. It reacts with metals, forming hydrides, and with the halogens (it combines violently with fluorine, even at very low temperatures).

About half of the hydrogen produced industrially is converted into ammonia. It is made by catalytic reduction (using nickel) of steam with hydrocarbon gases from oil refineries or with natural gas (methane):

$$C_nH_{2n+2} + nH_2O \rightarrow nCO + (2n + 1)H_2$$

and

$$C_nH_{2n+2} + 2nH_2O \rightarrow nCO_2 + (3n + 1)H_2$$

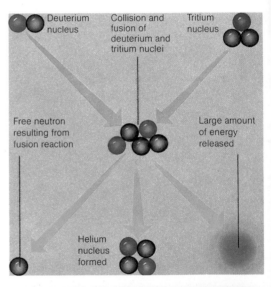

Deuterium nucleus — Collision and fusion of deuterium and tritium nuclei — Tritium nucleus

Free neutron resulting from fusion reaction — Large amount of energy released

Helium nucleus formed

A hydrogen bomb explosion (right) produces an enormous amount of energy virtually instantaneously. It works in two main stages: first an atomic (fission) bomb explodes to produce the extremely high pressure and temperature needed for the second stage, the fusion reaction. In this reaction (illustrated right, top), nuclei of deuterium and tritium fuse, producing a helium nucleus, a free neutron and a large amount of energy. Scientists are trying to develop fusion reactors in which the energy of fusion can be controlled and used to generate electricity.

This hydrogen is combined with nitrogen, by the Haber process, to produce synthetic ammonia:

$$3H_2 + N_2 \rightarrow 2NH_3$$

The ammonia is oxidized to nitrogen (II) oxide, which is then converted to nitric acid for making fertilizers and explosives.

Other important uses of hydrogen include the purification of oil refinery products (about 35 per cent of total use). It reduces the levels of contaminating sulfur, nitrogen and oxygen compounds, which interfere with the refining process. Many hydrogenation reactions, involving catalysts, are important in the production of fuels, and in fuel conversions — for example, the "upgrading" of heavy fuel oil to gasoline. Among many processes, the Fischer-Tropsch method is used for making hydrocarbons from coal and hydrogen, using various transition-metal catalysts. Hydrogen is used in the production of methanol, CH_3OH, by reacting it with carbon monoxide (572°F, or 300°C; 300 atmospheres pressure; zinc chromate catalyst). Both of these methods can be used for the production of liquid fuels from coal, but are uneconomic at present oil prices.

Other applications of hydrogen include the production of margarine by hydrogenation of animal, fish and vegetable fats and oils. It is also increasingly used for the direct reduction of iron ore to the metal.

Hydrogen isotopes

Heavy water, D_2O, the compound of deuterium and oxygen, occurs to the extent of about 1 part in 6,700 parts of "ordinary" water. It has a density 1.1059 times that of H_2O and a boiling point of 214.54°F (101.41°C). Some biological systems are affected by replacement of H_2O by D_2O — neither plants nor animals can survive indefinitely in D_2O. Heavy water is used in some nuclear reactors to moderate the behavior of neutrons.

The explosion of a hydrogen bomb involves the collision and fusion of small atomic nuclei, usually the isotopes deuterium and tritium. An atomic bomb is first exploded as a trigger to start the thermonuclear reaction of the hydrogen isotopes. The mixture of deuterium and tritium then reacts, producing a higher temperature, causing hydrogen fusion. The bomb also involves a fission reaction in a surrounding uranium material, which produces a wide range of radioactive isotopes as fission products. If these reactions could be controlled (the "fusion" reactor), then the deuterium of heavy water could be a virtually unlimited source of energy.

The *Hindenburg*, a hydrogen-filled airship 800 feet (245 meters) long, burst into flames on approaching the mooring mast at Lakeside, New Jersey, on May 6, 1937. As a result of this still unexplained disaster interest in airships waned, although it has recently revived and several experimental airships have been built. These modern craft, however, are filled with helium, which is not inflammable.

Liquid hydrogen has been used as a propellant for several spacecraft, including the second and third stages of the Saturn V rocket (left) used to launch the historic Apollo 11 mission that landed the first men on the Moon. In such propulsion systems, the liquid hydrogen is reacted with liquid oxygen, producing water (in the form of steam) — and a powerful thrust.

Fact entries

Hydrogen was first distinguished from other inflammable gases in 1766 by the British chemist Henry Cavendish (1731—1810). The formation of water by burning the gas was observed in 1776, and the name hydrogen (from the Greek *hydro* and *genes*, meaning "water former") was proposed in 1781 by the French chemist Antoine Lavoisier (1743—1794). At. no. 1; at. mass 1.008; m.p. —259.2°C; b.p. —252.8°C. Isotopes: ^1H (protium; relative abundance 99.98%); ^2H (deuterium, see below); ^3H (tritium, see below).

Deuterium, the isotope of hydrogen with a mass of about 2, was discovered in 1931 by the American physicist Harold Urey (1893—1981). At. no. 1; at. mass 2.014; m.p. —254.4°C; b.p. —249.5°C; relative abundance 0.0156%. made by electrolysis of heavy water.

Tritium, the isotope of hydrogen with a mass of about 3, was discovered in 1934 by the British physicist Ernest Rutherford (1871—1937) and the Australian physicist Marcus Oliphant (1901—   ). At. no. 3; at. mass 3.022; m.p. —252.5°C; b. p. —248.1°C; relative abundance about $10^{-15}\%$; radioactive (half-life 12.26 years).

1		
H		
1.008		
IA	IIA	
3	4	
Li	Be	
6.941	9.012	
11	12	
Na	Mg	
22.990	24.305	
19	20	
K	Ca	
39.102	40.080	
37	38	
Rb	Sr	
85.468	87.620	
55	56	
Cs	Ba	
132.906	137.340	
87	88	
Fr	Ra	
(223)	(226)	

The alkali metals constitute Group IA of the Periodic Table. In general, each alkali metal element is more reactive than is its counterpart in the neighboring Group IIA.

Alkali metals

The six elements of Group IA of the Periodic Table are known as the alkali metals. They are lithium (Li), sodium (Na), potassium (K), rubidium (Rb), cesium (Cs) and francium (Fr). All except lithium are highly reactive, burn in air, and react vigorously — sometimes explosively — with water to form strongly alkaline hydroxides. Because of their reactivity the free metals (again except lithium) are normally stored out of contact with air, usually under a layer of hydrocarbon oil. Sodium and potassium occur widely in nature as salts, and are essential for many forms of life, whereas francium, on the other hand, is one of the rarest naturally occurring elements. Sodium and potassium are by far the most important alkali metals industrially.

Lithium
The first member of the series, lithium, is non-typical in its properties — in some ways it is more like an alkaline earth metal such as magnesium or calcium. Its usual source is the mineral spodumene, lithium aluminum silicate (which contains about 3.7 per cent lithium). Spodumene is converted to the lithium sulfate (Li_2SO_4) and then the chloride (LiCl), which is electrolyzed to give the metal.

Lithium is the lightest metal of all (relative density 0.534). It is used in various alloys, notably with copper, and as a degasifier in producing "hard" vacuums. Lithium soaps (such as the stearate, $C_{17}H_{35}COOLi$) are used in lubricating greases. Lithium salts are used in the treatment of some types of mental disorders.

Sodium and potassium
Sodium is the sixth most abundant element, occurring mainly as sodium chloride (NaCl, common salt), the major inorganic component of seawater. In some tropical countries, evaporation of seawater using the heat of the Sun has been used for many centuries as a production method. In other parts of the world, underground salt domes — usually "mined" by leaching with water under pressure — are the principal source of salt. Other natural sources of sodium include soda ash (sodium carbonate, Na_2CO_3) and borax (sodium perborate, $Na_2B_4O_7$).

Much of the sodium chloride produced is converted by electrolysis to sodium hydroxide (NaOH, caustic soda) and chlorine, both of which have many industrial applications. Soda ash is a basic raw material for the manufacture of glass and some ceramics. Other widely used sodium salts include the tripolyphosphate, tetraborate and dodecylbenzene sulfonate (all used in household detergents), chlorate ($NaClO_3$, a weed-killer), dichromate ($Na_2Cr_2O_7$, used in photographic emulsions), sulfide (Na_2S) and hydrosulfide (NaSH), for leather tanning, and the thiosulfate ($Na_2S_2O_3$, a fixer in photographic processing). Metallic sodium, also obtained by electrolysis, is used as a coolant in some nuclear reactors and a sodium-lead alloy is employed in making lead tetraethyl, $Pb(C_2H_5)_4$, the form in which lead is used in the much-disputed anti-knock additive to gasoline.

Potassium is also an abundant element, but occurs mainly in rocks and clays and is difficult to extract. Seawater contains potassium chloride (KCl), but the main economic sources are dried salt beds and the Dead Sea. Large amounts of impure potassium chloride are mined for use as fertilizers, because potassium is an essential element for plant growth, and is found in all soils. Potassium salts are more expensive than those of sodium, but because they are not hygroscopic

The electrolysis of brine — essentially a solution of sodium chloride — in a diaphragm cell (illustrated right) is one of the principal methods of obtaining sodium hydroxide and chlorine, both of which are important industrial raw materials. The details of the electrolysis are shown in the diagram, but overall, brine enters the cell, is depleted of chloride ions in the anode compartment (yielding chlorine gas) and of hydrogen ions in the cathode compartment (giving hydrogen gas) to produce a caustic liquid. This liquid is drained from the cell and evaporated to remove the remaining chloride ions (which crystallize out as sodium chloride), leaving sodium hydroxide solution.

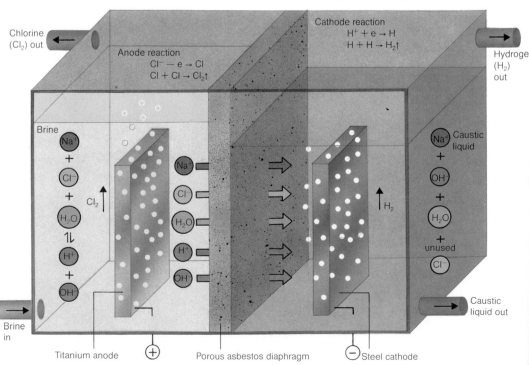

Chlorine (Cl_2) out

Anode reaction
$Cl^- - e \rightarrow Cl$
$Cl + Cl \rightarrow Cl_2\uparrow$

Cathode reaction
$H^+ + e \rightarrow H$
$H + H \rightarrow H_2\uparrow$

Hydrogen (H_2) out

Brine

Na$^+$
Cl$^-$
H_2O
H^+
OH$^-$

Cl_2
H_2

Na$^+$ Caustic liquid
OH$^-$
H_2O
unused
Cl$^-$

Caustic liquid out

Brine in

Titanium anode (+)

Porous asbestos diaphragm

Steel cathode (−)

(water-attracting), they are employed where damp resistance is important. Examples include the use of potassium nitrate (KNO_3, saltpeter) in gunpowder and fireworks, and potassium chlorate ($KClO_4$) in match-heads. In higher animals, potassium ions (K^+), with sodium ions (Na^+), act at cell membranes in transmitting electrochemical impulses in nerve and muscle fibers and in balancing the activity of food intake and waste removal from cells. This metabolic process, known as the Na^+/K^+ ion pump, accounts for the transport of K^+ ions in and Na^+ ions out through the living cell membrane. This pump is present in all cells of animals, plants and microorganisms. Details of the process are still unknown; however, too little or too much potassium in the body is fatal.

Potassium is slightly radioactive, because of the presence of the unstable isotope ^{40}K (half-life 1.26×10^9 years), which decays at a known rate to an isotope of argon (^{40}Ar). The rate can be measured, and is used for determining the ages of certain igneous and metamorphic rocks, such as micas. Potassium-argon dating is subject to some inaccuracy (because of the loss of argon gas by diffusion), but it is a useful method within its limitations.

Rubidium, cesium and francium
Rubidium is widely distributed in the Earth's crust with other alkali metals (especially cesium) in small amounts, although it is actually more abundant overall than lead, copper or zinc. It is a reactive metal, like cesium, but has few uses. Cesium is used as the time-measuring element in atomic clocks. Chemically, its hydroxide (CsOH) is the strongest known base, attacking even glass.

Natural rubidium contains nearly 28 per cent of the radioactive isotope ^{87}Rb (half-life approximately 5×10^{10} years), which decays to an isotope of strontium, ^{87}Sr. The ages of meteorites can be estimated by measuring their ratio of ^{87}Rb to ^{87}Sr. One such estimate for a stony meteorite puts the age of the Solar System at about 4,500 million years.

Francium has more than 20 isotopes, all of which are unstable and radioactive. ^{223}Fr, a breakdown product of actinium, has the longest half-life (21 minutes). No weighable amount of the element has even been extracted, and at any one time there is only about 25 grams of it throughout the whole of the Earth's crust.

Sodium vapor is used in some streetlights because the orange light emitted when an electric current passes through the vapor is not only intense but also penetrates mist and fog better than does white light.

Many types of preserved meats — most sausages, for example — contain potassium nitrate (KNO_3) because it inhibits the growth of harmful microorganisms such as *Salmonella* (which causes food poisoning).

Fact entries

Lithium was discovered in 1817 by the Swedish chemist Johan Arfvedson (1792—1841). Its name derives from the Greek *lithos,* meaning stone. At. no. 3; at. mass 6.941; m.p. 180.5°C; b.p. 1,336°C.

Sodium was discovered in 1807 by the British scientist Humphry Davy (1778—1827), who named it after the alkaline substance soda. At. no. 11; at. mass 22.990; m.p. 97.8°C; b.p. 881.4°C. The symbol for sodium, Na, comes from its Latin name *natrium.*

Potassium (originally called kalium, the Latinized form of the Arabic word for alkali) was the first metal to be isolated by electrolysis, by Humphry Davy in 1807.

At. no. 19; at. mass 39.102; m.p. 63.2°C; b.p. 765.5°C.

Rubidium was discovered spectroscopically in 1861 by the German chemists Robert Bunsen (1811—1899) and Gustav Kirchhoff (1824—1887). It is so named because of the dominant red lines in its spectrum (the Latin *rubidus* means red). At. no. 37; at.

mass 85.468; m.p. 39°C; b.p. 688°C.

Cesium was discovered spectroscopically in 1860 by Robert Bunsen and Gustav Kirchhoff. It was named after the characteristic blue lines in its spectrum (the Latin *caesius* means gray blue). At. no. 55; at. mass 132.906; m.p. 28.5°C; b.p. 705°C.

Francium was discovered (as a transitory decay product of actinium) in 1939 by the French chemist Marguerite Perey (1900—1975), who named it after her homeland. Little is known about the element because no stable isotopes exist; ^{223}Fr, the only natural isotope, is the longest-lived (half-life 21 minutes).

VIII	IB	IIB
28 **Ni** 58.700	29 **Cu** 65.546	30 **Zn** 65.380
46 **Pd** 106.400	47 **Ag** 107.868	48 **Cd** 112.400
78 **Pt** 195.090	79 **Au** 196.967	80 **Hg** 200.590

Copper, silver and gold constitute Group IB of the Periodic Table. All are malleable, ductile metals with high electrical conductivities.

Copper, silver and gold

Sometimes called the coinage metals, copper (Cu), silver (Ag) and gold (Au) make up Group IB of the Periodic Table. Unlike the alkali metals of Group IA, they do not display any regularity of properties. Each metal does, however, have a range of valuable features, both individually and as alloys and compounds.

Copper and its compounds
The least intrinsically valuable of the three, copper has been known for about 10,000 years. It is the principal nonferrous metal, and one of the few common metals which find their main uses in the pure form — rather than as alloys — especially in electrical applications. The sulfide ores are widely distributed, especially in North and South America, and in central Africa; many of the ores contain less than one per cent of copper. The ore is crushed, concentrated by flotation, and smelted to give the impure metal, which is purified by electrolysis (often yielding nickel, silver and other precious metals in the sludge). Electrolytic purification, although expensive, is essential to produce pure copper with the high conductivity required for electrical uses.

Copper exhibits two oxidation states and both copper (I) (Cu^+) and copper (II) (Cu^{2+}) compounds are common. It has good corrosion resistance — the characteristic green coating seen on the weathered metal is a protective film of hydrated copper carbonate, with basic copper sulfate in industrial atmospheres. The metal forms many complexes (coordination compounds), and in this form is a catalyst for several important organic reactions. Some of these complexes are used in electroplating solutions — copper is often electroplated as an undercoat on steel, before plating with nickel or chromium, to improve adhesion and wear. Important copper alloys include cupronickel (for coins), brass (copper with up to 20 per cent zinc) and various bronzes (copper with phosphorus, zinc, tin, manganese or other metals).

Copper plays a key part in two main areas of plant metabolism: the synthesis of the photosynthetic green pigment chlorophyll, and the activity of enzymes, for which its presence is sometimes essential. It occurs in enzymes which catalyze various oxidation-reduction reactions. In animals, copper is also essential to metabolism, especially in the blood (human adults need about 2 milligrams per day), but large amounts are toxic. Copper sulfate ($CuSO_4$) in Bordeaux mixture is a traditional fungicide.

Silver
Silver was a precious and sacred metal in many ancient civilizations, second only to gold. Silver sulfide (Ag_2S), the most common ore, occurs as a minor proportion with those of copper and lead; it is usually obtained as a processing by-product. Slag dumps in Asia Minor and the Greek Islands show that people had learned to separate silver from lead by 4000BC. Pure silver is highly ductile and malleable, with the highest thermal and electrical conductivity and lowest contact resistance of all metals. It does not oxidize in air, but the molten metal readily absorbs oxygen. It readily combines with sulfurous gases in the air, becoming tarnished — the dark brown film can be removed by polishing or chemical solution. Silver (like gold) is attacked by cyanide solutions, which are the basis of its usual electroplating solutions.

Silver is widely used in photographic emulsions (about 30 per cent of its total use) in the form of silver nitrate, which is converted to halide crystals — mainly silver bromide with some iodide. The halides are dispersed in gelatin and coated on glass or a transparent plastic film (usually cellulose acetate). The halide crystals are

Two micrographs show the silver halide crystals — magnified about 3,750 times — in conventional photographic film (near right) and in a new ultra-sensitive film (far right). The new film is more sensitive because the flat, regular shape of its large silver halide crystals enables them to absorb more light than the irregular, boulderlike crystals of the other film.

pale yellow in color. They darken on exposure to light, turning black as metallic silver is formed. The addition of a reducing agent (the "developer") causes crystals exposed to light to be totally converted to silver, forming the negative photographic image. Unaffected crystals are removed by washing with a "fixer" — usually sodium thiosulfate solution — which removes the silver salt as a soluble complex. Photographic emulsions contain about $1/2$ ounce (15 grams) of silver per square yard (square meter); with increasing scarcity (and cost) of the metal, recovery and recycling techniques for scrap film and developer (and electroplating solutions) are now well established.

Silver is being increasingly used in batteries, both primary (single-use) and secondary (rechargeable) types. The former have silver chloride and magnesium electrodes, with a salt electrolyte; the latter include silver-cadmium and silver-zinc cells, both of which are sealed with a potassium hydroxide electrolyte. Silver is also used in many alloys, most often with copper. They include "sterling silver" (Ag:Cu ratio 92.5:7.5) and "coin silver" (Ag:Cu ratio 90:10), which is no longer used for coins but employed for electrical contacts.

Gold

Gold has been mined, used and treasured by mankind since prehistoric times, and gold coins, jewelry and emblems form an important part of recorded history. The first true coins (of standard weight and value) were probably produced in Anatolia in about 650BC from a natural alloy of gold containing 20—30 per cent of silver. It is widely distributed in small amounts, usually as the metal, which is recovered from the ore by smelting, by amalgamation (dissolving in mercury) or by treatment with alkaline cyanide solution, from which it is extracted by electrolysis. The pure metal, which has a high density (19.32), is the most ductile and malleable of all metals, and is the third best conductor of heat and electricity (after silver and copper). Gold leaf, made by carefully beating the metal, can be as little as 4 millionths of an inch (10^{-5}cm) thick. It transmits green

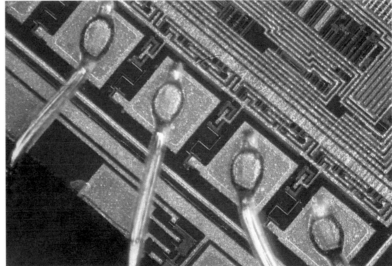

light, but appears golden in reflected light. Gold is extremely soft, and so is usually alloyed with copper, silver, zinc or nickel for practical applications. The proportion of gold in alloys is often expressed in carats. Pure gold is 24 carats; therefore 18-carat gold is 18/24 or 75 per cent gold.

Apart from coins, jewelry and dentistry, the metal is increasingly employed in electroplated contacts for electronic circuits: although expensive, gold contacts give improved reliability. As a "noble" metal, it is inert to most acids and alkalis, but dissolves in aqua regia (a 4:1 mixture of hydrochloric and nitric acids).

The electrical contacts of microchips (above) and other electronic devices are often plated with gold, which is not only a good electrical conductor, but also protects the contacts from corrosion and therefore improves reliability.

Fact entries

Copper was first used by late Stone-Age man around 8000BC. Its name derives from the Latin word for the metal, *cuprum*, from which its chemical symbol (Cu) also comes. It occurs naturally as the free metal in basaltic lavas and also as copper compounds in many minerals. At. no. 29; at. mass 63.546; m.p. 1,083°C; b.p. 2,595°C.

Silver has been known since prehistoric times; silver ornaments dating from about 4000BC have been found in royal tombs. Its chemical symbol, Ag, is derived from the Latin word for the metal, *argentum*. It occurs naturally in the free metallic state and as compounds in various minerals. At. no. 47; at. mass 107.868; m.p. 960.5°C; b.p. about 2,000°C.

Gold, like copper and silver, has been known since prehistoric times; because it does not corrode and mainly occurs naturally in a relatively pure form, it was one of the first metals used by people. Its chemical symbol, Au, comes from the Latin word for the metal, *aurum*. Gold exhibits two oxidation states: Au (I) and Au (III). At. no. 79; at. mass 196.967; m.p. 1,064.8°C; b.p. 2,700°C.

H 1.008		
IA	IIA	
3 Li 6.941	4 Be 9.012	
11 Na 22.990	12 Mg 24.305	IIIB
19 K 39.102	20 Ca 40.080	21 Sc 44.956
37 Rb 85.468	38 Sr 87.620	39 Y 88.906
55 Cs 132.906	56 Ba 137.340	57 La 138.906
87 Fr (223)	88 Ra (226)	89 Ac (227)

The alkaline earth elements make up Group IIA of the Periodic Table. Chemically they resemble, but are less reactive than, their counterparts in Group IA but have little similarity to their other neighbors in Group IIIB.

Alkaline earths

The six elements of Group IIA of the Periodic Table are known as the alkaline earth metals; the series is beryllium (Be), magnesium (Mg), calcium (Ca), strontium (Sr), barium (Ba) and radium (Ra). Their oxides are the actual alkaline earths, so called because of their basic nature. All of the metals react with water and, except for radium, are of industrial importance. They form a closely related group of highly metallic elements with a regular gradation of properties down the group. All have oxidation state II in their many compounds.

Beryllium and magnesium
Beryllium is much less common than would be expected theoretically because its atoms are disrupted by natural bombardment into helium gas. The pure metal, usually made by heating beryllium fluoride (BeF_2) with magnesium metal, has a low neutron absorption and so is used as a "moderator" in nuclear reactors. It slows fast neutrons, making it possible to control the fission process. It is strong for its weight and ductile, hence its use in aerospace components. Beryllium compounds are highly poisonous; for this reason, the former use of the beryllium oxide (BeO) as the internal coating of fluorescent lamps has largely been discontinued, because of the problems of safe disposal.

Magnesium is the lightest metal in general engineering use — mainly as alloys containing up to 10 per cent of other metals (usually aluminum or zinc) to increase its strength. The metal itself is employed in incendiary bombs and other pyrotechnic devices. The sulfate ($MgSO_4$, Epsom salts) and the hydroxide ($Mg(OH)_2$, "milk of magnesia") are both used as laxatives and purgatives. The French chemist Victor Grignard (1871—1935) found that magnesium readily combines with organic halides, giving organomagnesium compounds that are widely used in synthesizing complex organic substances.

Calcium, strontium and barium
Calcium is the fifth most abundant element in the Earth's crust and the most common (and cheapest) alkaline earth element, in both occurrence and use. It occurs naturally in a wide range of forms: as calcium carbonate ($CaCO_3$, chalk, marble, limestone and the shells of mollusks and crustaceans), calcium sulfate (anhydrite, $CaSO_4$, and gypsum, $CaSO_4.2H_2O$), calcium fluoride (CaF_2, fluorite) and calcium phosphate ($Ca_3(PO_4)_2$, in bones and teeth). Calcium metal, made either by electrolysis of the molten chloride ($CaCl_2$) or by the reduction of lime (CaO) with aluminum at about 2,200°F (1,200°C), is attacked by moist air, giving a coating of the hydroxide ($Ca(OH)_2$) and evolving hydrogen.

Immense quantities of calcium compounds are used in building and construction. For many cen-

Several alkaline earth elements are used in fireworks (left), flares and other pyrotechnic devices because they burn easily and produce an intense light. Moreover, each element (and its compounds) emits light of a characteristic color; for example, the white colors in the firework display were produced by magnesium, the red colors by strontium, and the green by barium.

The shells of many mollusks — such as the helmet cowrie (Cypraecassis rufa) in the photograph below — and crustaceans consist largely of calcium carbonate, which is secreted by the underlying tissues to form the shell. Calcium salts also play an important role in other animals — helping to form bones and teeth in humans and other vertebrates, for example.

turies limestone ($CaCO_3$) has been heated to give carbon dioxide (CO_2) and quicklime (CaO, calcium oxide), which is converted with water to slaked lime ($Ca(OH)_2$, calcium hydroxide), used in mortar and brickwork. Calcium compounds are also the basis of cement and plasters, with complex setting mechanisms which control the application and final properties of the products.

Calcium compounds are also essential in living organisms. In animal tissues, calcium ions help to control nerve-impulse transmission, muscle action and blood clotting. Calcium deficiency causes rickets and failure of blood clotting, whereas excessive calcium absorption causes calcification of joints and kidney stones (calculi).

Strontium compounds have few industrial applications, because cheaper calcium or barium compounds can generally be used instead. The artificial isotope ^{90}Sr (half-life 28 years), which occurs in the debris of atomic explosions, is very hazardous, because it can be ingested by humans and deposited in the bones — replacing calcium — and emitting beta radiation, which damages the bone marrow.

Barium compounds are obtained from the mineral barytes ($BaSO_4$, barium sulfate), one of the least soluble substances known. It is used as a dense filler for paints and paper, and as the radiopaque medium in internal X-ray examinations. Soluble barium salts, such as barium chloride ($BaCl_2$), are highly poisonous.

Radium

Radium and its compounds are radioactive — the property used by Pierre and Marie Curie in 1898 to follow the separation of radium compounds, with barium, from uranium ores. Radium is usually used — as the sulfate ($RaSO_4$) — for its radioactivity: neutron sources are made from $RaSO_4$ and beryllium metal powder, and the salt is an ingredient in some luminous paints. Radium compounds are the chief source of the rare gas radon. The curie, a common unit of radioactivity, was originally defined as the amount of radioactivity equal to that produced by one gram of the isotope ^{226}Ra but is now defined as 3.7×10^{10} disintegrations per second.

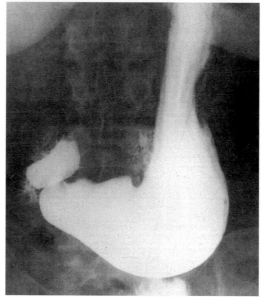

Hang gliders (above) often use magnesium-aluminum alloy for their frames because of its combination of light weight and strength.

The X-ray photograph (left) shows a human stomach, which is visible only because the patient has swallowed a "barium meal" containing barium sulfate, a substance that is opaque to X rays.

Fact entries

Beryllium was discovered as its oxide in the mineral beryl (a silicate of beryllium, from which its name derives) in 1798 by the French chemist Louis Vauquelin (1763—1829). The free metal was first isolated in 1828 by the German chemist Frederick Wöhler (1800—1882) and, independently, by the French chemist Antoine Bussy (1794—1882). At. no. 4; at. mass 9.012; m.p. 1,287°C; b.p. about 2,500°C.

Magnesium was first isolated as the free metal in 1808 by the British scientist Humphry Davy (1778—1827), although its compounds (Epsom salts and magnesia, for example) were known much earlier. The element is named after the ancient Greek region of Magnesia. At. no. 12; at. mass 24.305; m.p. 651°C; b.p. 1,100°C.

Calcium was first isolated as the free metal in 1808 by Humphry Davy. Its name is derived from the Latin *calx*, meaning lime — a calcium compound known since ancient times. At. no. 20; at. mass 40.080; m.p. 850°C; b.p. 1,440°C.

Strontium was first found as the carbonate ($SrCO_3$) at Strontian (hence the element's name) in Scotland in 1787 by the British scientist William Cruikshank. The metal itself was isolated in 1808 by Humphry Davy.

At. no. 38; at. mass 87.620; m.p. 757°C; b.p. 1,366°C.

Barium derives its name from the Greek *barys*, meaning heavy — although the metal is not particularly dense. It was first detected in 1774 by the Swedish chemist Karl Scheele (1742—1786), although the free metal was not isolated until 1808, by Humphry Davy. At. no. 56; at. mass 137.340; m.p. about 710°C; b.p. about 1,600°C.

Radium was discovered in the radioactive mineral pitchblende (a uranium ore) in 1898 by the French scientists Pierre Curie (1859—1906) and his wife Marie (1867—1934). The metal itself was isolated in 1910 by Marie Curie. Its name is derived from the Latin *radius*, meaning ray. At. no. 88; at. mass (of stablest isotope) 226; m.p. about 700°C; b.p. about 1,737°C; half-life of ^{226}Ra 1,600 years.

IB	IIB	IIIA
		5 **B** 10.810
		13 **Al** 26.982
29 **Cu** 63.546	30 **Zn** 65.380	31 **Ga** 69.720
47 **Ag** 107.868	48 **Cd** 112.400	49 **In** 114.820
79 **Au** 196.967	80 **Hg** 200.590	81 **Tl** 204.370

Zinc, cadmium and mercury constitute Group IIB of the Periodic Table. They are chemically similar (all forming compounds in oxidation state II, for example), in contrast to their physical characteristics.

Zinc, cadmium and mercury

These three metals are classed together as Group IIB of the Periodic Table. They are superficially dissimilar: zinc (Zn) is a soft metal, cadmium (Cd) and its compounds are poisonous, and mercury (Hg) is the only metallic element that is liquid at ordinary temperatures. But chemically they behave alike. All form compounds in oxidation state II, although mercury also forms some compounds in oxidation state I.

Zinc

Zinc was first smelted in China and India, in about AD 1000, and the technology appeared in the West in the eighteenth century when the metal was called Indian tin or calamine (a term now reserved for natural zinc carbonate). Zinc is comparatively rare in nature, although widely distributed, occurring mainly as the sulfide (ZnS, zinc blende) to the extent of about 4 ounces per ton (120 grams per metric ton) in the Earth's crust. But it is the fourth most common metal in industrial use (after iron, copper and aluminum). The sulfide ore is roasted in air to the crude oxide (ZnO), which is reduced with carbon and the resultant metal vapor condensed to the solid metal, which is cast into ingots.

Little zinc is used alone, but large amounts are employed in alloys, notably in die-casting (with aluminum, and a little copper and magnesium), brass (which is 3 to 45 per cent zinc, with copper), and in galvanizing, in which steel is hot-dipped in molten zinc and so protected against corrosion by a thin zinc skin. When the protective coating is damaged, the protection is continued by electrochemical action, as the Zn is slowly consumed. This "sacrificial" action is used in the cathodic protection of ships and buried steel structures, where an electrical circuit is deliberately set up between the steel to be protected and zinc cathodes, which are allowed to dissolve.

Chemically, zinc forms many colorless compounds, which have considerable industrial importance. The oxide (ZnO) is a paint pigment (zinc white), the sulfate ($ZnSO_4$) is used in making viscose rayon and in the froth flotation of minerals, and zinc chloride ($ZnCl_2$) is employed as a soldering flux, in dry batteries, and in embalming.

Zinc is an essential trace element in the human body in red blood cells, where it plays an important part in carbon dioxide metabolism. Its presence in the pancreas may be necessary for insulin storage. In plants, zinc deficiency can cause leaf disease in trees.

Cadmium

Cadmium is obtained as a by-product from zinc, lead and copper ores, and was first discovered as an impurity in a type of zinc carbonate. The metal is mainly used as a protective plating on steel and other metals, in many alloys — to improve the strength and ductility of copper for electrical contacts and terminals — and in many types of bearings, especially for high-temperature use. It is also a component of nickel-cadmium rechargeable batteries, with potassium hydroxide as the electrolyte. In addition, cadmium is used extensively in electronics, in rectifiers and phosphors in television screens, and — as the sulfide (CdS) — in photoelectric devices and in solar cells.

Cadmium forms an oxide (CdO) in moist air and, like the other Group IIB elements, is converted by acids to the corresponding salts, such as the nitrate, $Cd(NO_3)_2$, and the cyanide $Cd(CN)_2$. Most of the salts, and the oxide, are dark red in color; some cadmium sulfides are used as orange-yellow pigments. Cadmium salts of organic acids are widely used as stabilizers for polyvinyl chloride (PVC) plastic, in which they reduce yellowing. The element is highly toxic, especially as inhaled dust. Deaths from acute cadmium poisoning have occurred from inhalation of CdO fumes, usually from the welding of cad-

The suspended steel pipes in the photograph (right) have just been galvanized — coated with a layer of zinc by dipping them into a bath of the molten metal. The zinc coating on galvanized steel prevents the steel from rusting; the zinc itself does not corrode easily because, in air, a thin film of zinc oxide forms on its surface, thereby protecting the underlying metal.

The ancient rock paintings (left) show the characteristic red color of cinnabar, a naturally occurring sulfide ore of mercury.

mium-plated steel without adequate ventilation. Most cadmium in the environment results from effluent from plating processes, which can poison biological sewage treatment plants.

Mercury

Also called quicksilver, mercury is a dense silvery metal, and is the only elemental metal that is liquid at room temperature. It usually occurs in nature in small amounts as the sulfide (HgS). Red mercury sulfide, which occurs naturally in the ore cinnabar, is still sometimes used to make the pigment vermilion, although its use is limited by its toxicity. The chief producers are Spain, Italy and the Soviet Union; world production is less than 11,000 tons (10,000 metric tons) per year. The ore is converted by heat to the metal.

Mercury is mainly used in long-life batteries, with good electrical stability, for radios, watches, cameras and calculators. Mercury vapor is employed in electrical discharge lamps, producing a high intensity bluish light rich in ultraviolet radiation, and in fluorescent tubes. The metal is also used in industrial cells for the electrolytic production of sodium hydroxide and chlorine. It forms alloys called amalgams, some of which are used for dental fillings. Fine particles of native gold and silver are extracted by combination with mercury, which is recovered by distillation.

Mercury is toxic, and poisoning may result from inhaling the vapor, absorption through the skin, or ingestion of soluble compounds. A his-

torical example is hatter's disease, contracted by workers from the use of mercury compounds in preparing felt from rabbit fur. For this reason, the former important uses of mercury compounds in fungicides and antimicrobial mixtures are now strictly limited. One compound, mercury fulminate, $Hg(ONC)_2$, has been employed as a detonating explosive ever since its discovery in 1799.

Cadmium sulfide occurs in several forms, the yellow variety being employed as a pigment (called cadmium yellow) in paints, such as that used on the earthmovers (below).

Fact entries

Zinc has been known under various names for at least 2,500 years, usually as an alloy. The free metal was isolated in about the eleventh century AD, by early Indian and Chinese metallurgists. At. no. 30; at. mass 65.380; m.p. 419.5°C; b.p. 908°C.

Cadmium was discovered in 1817 as an impurity in a type of zinc carbonate called cadmia (from which the element derives its name) by the German chemist Friedrich Strohmeyer (1776–1835). At. no. 48; at. mass 112.400; m.p. 321°C; b.p. 765°C.

Mercury was named after the Roman god Mercury; its symbol, Hg, is derived from the Latin hydrargyrum, meaning liquid silver. The element has been known since ancient times; samples have been found in tombs dating from about 1500BC. At. no. 80; at. mass 200.590; m.p. −38.87°C; b.p. 356.72°C. This range of temperature makes it useful in thermometers.

The boron group

Group IIIA of the Periodic Table comprises the boron group of elements: boron (B), aluminum (Al), gallium (Ga), indium (In) and thallium (Tl). Each element has three electrons in its outer shell, and as a result most of their ionic compounds have oxidation states of three, although the lower members form different types of bonds in certain compounds. All except boron are metals.

Boron

Boron was isolated in 1808, although its principal compound, borax ($Na_2B_4O_7.10H_2O$), has been known since Babylonian times as a mild antiseptic, detergent, and water-softening material. Boron as an element has some specialized uses as a high-stiffness fiber in aerospace applications, and as a thin film in nuclear devices. Its chemistry is complex and unusual: it forms compounds with the halogens, of which the most notable is the corrosive toxic gas boron trifluoride (BF_3) — used as an acid catalyst in many organic reactions, such as synthetic rubber production. Compounds of boron with other elements are outstandingly stable (with melting points above 3,600°F, or

2,000°C) and hard — boron nitride ("borazon") and aluminum boride can be used as substitutes for diamond in metal grinding and polishing. Borosilicate glass has a low coefficient of thermal expansion, and is used wherever heat- and shock-resistance is important.

Trace amounts of boron are essential for good plant growth; "brown heart" of vegetables and "dry rot" of sugar beet are caused by boron deficiency, and can be treated by addition of soluble borates to the soil. Large amounts of boron compounds are, however, poisonous.

Unlike the rest of this group of elements, boron readily forms compounds resembling those of carbon and silicon, notably the boron hydrides. These have unusual electron-deficient bonds, which have an important place in the theory of chemical bonding. The boron hydrides (which have been tried as rocket fuels) include many compounds with complex cagelike, three-dimensional structures. They are obtained by reducing boron halides with lithium aluminium hydride ($LiAlH_4$) to diborane (B_2H_6, the simple BH_3 does not normally exist), which can be converted by various methods such as heat or electric discharge into the more complex boranes.

The boron group of elements makes up Group IIIA of the Periodic Table. Boron and aluminum are the most important members; the remaining three have few uses.

	IIIA	IVA
	5 B 10.810	6 C 12.011
IIB	13 Al 26.982	14 Si 28.086
30 Zn 65.380	31 Ga 69.720	32 Ge 72.590
48 Cd 112.400	49 In 114.820	50 Sn 118.690
80 Hg 200.590	81 Tl 204.370	82 Pb 207.200

The principal stages in the industrial extraction of aluminum from its chief ore, bauxite, are shown in the flow diagram (right). The key stage is the electrolysis of molten alumina and cryolite (the details of which are illustrated in the diagram bottom right), which produces molten, pure aluminum. This process requires a large amount of electricity, so aluminum extraction plants (below) are often sited in areas where electricity is cheap — near hydroelectric power stations, for example.

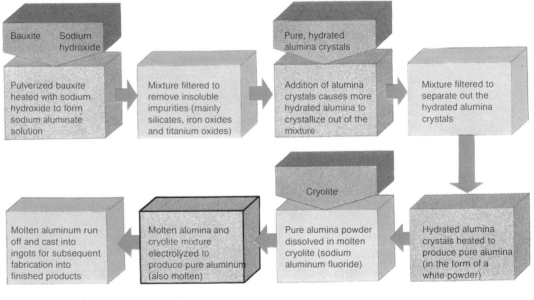

Bauxite   Sodium hydroxide

Pulverized bauxite heated with sodium hydroxide to form sodium aluminate solution

Mixture filtered to remove insoluble impurities (mainly silicates, iron oxides and titanium oxides)

Pure, hydrated alumina crystals

Addition of alumina crystals causes more hydrated alumina to crystallize out of the mixture

Mixture filtered to separate out the hydrated alumina crystals

Molten aluminum run off and cast into ingots for subsequent fabrication into finished products

Molten alumina and cryolite mixture electrolyzed to produce pure aluminum (also molten)

Cryolite

Pure alumina powder dissolved in molten cryolite (sodium aluminum fluoride)

Hydrated alumina crystals heated to produce pure alumina (in the form of a white powder)

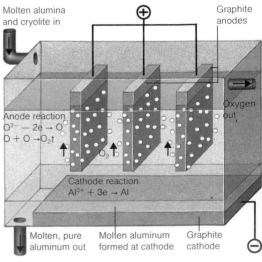

Molten alumina and cryolite in

Graphite anodes

Anode reaction
$O^{2-} - 2e \rightarrow O$
$O + O \rightarrow O_2\uparrow$

Oxygen out

Cathode reaction
$Al^{3+} + 3e \rightarrow Al$

Molten, pure aluminum out

Molten aluminum formed at cathode

Graphite cathode

Aluminum

Aluminum is the most abundant metallic element — making up about 8.3 per cent of the Earth's crust — and the third most abundant (after oxygen and silicon) of all elements. It is a component of most clays, which are aluminosilicates, but the most important ore is bauxite (named after Les Baux, in France, where it was first obtained), an impure aluminum oxide (Al_2O_3). In producing the metal, bauxite is first refined to alumina (pure aluminum oxide), which is smelted with cryolite (sodium aluminum fluoride) in an electric arc process, which has remained basically unchanged since it was invented independently in 1886 by the American Charles Hall (1863—1914) and the Frenchman Paul Héroult (1863—1914). Enormous effort has been devoted to saving electricity in this energy-intensive process, which uses 13 to 16 kWH to produce each kilogram of metal.

Aluminum is probably not essential for plant and animal life, but accumulates in certain species. The blue color of hydrangea flowers and some fruits is caused by an aluminum anthocyanin compound. The blue color can sometimes be stimulated by addition to the soil of alum (a term which includes several compounds — the most common is potash alum, the double salt K_2SO_4. $Al_2(SO_4)_3$.$24H_2O$).

Gallium, indium and thallium

The existence of gallium was predicted by Dmitri Mendeleyev in his original Periodic Table, and was subsequently discovered in 1875 by the Frenchman Lecoq de Boisbaudran.

Gallium is similar to aluminum in its chemical properties, but is rather less reactive. It occurs as a very minor component (0.005 to 0.01 per cent) in bauxite and zinc ores; it is separated from their residues and obtained as the metal by electrolysis. Most gallium is used in semiconductors, in very pure form (greater than 99.9999 per cent) achieved by electrochemical deposition. World demand is about 30 tons (27 metric tons), per year, mostly as gallium arsenide and gallium phosphide (for light-emitting diodes) and as gadolinium-gallium garnets (for computer memories).

Indium is a soft, lustrous, low-melting metal, rare in the Earth's crust; it is usually obtained as a by-product from concentrating zinc ores. Formerly used with germanium in semiconductors (now replaced by silicon), its main applications are in heavy-duty bearings for diesel engines and in low-melting solders.

Thallium is also a rare element. It is obtained from the flue dust from smelting lead and zinc, the ores of which contain small amounts of thallium. It forms thallium (I) and thallium (III) salts (with oxidation states of one and three), which can be readily interconverted. Thallium and its salts are highly toxic; poisoning, which can be caused by skin contact alone, causes rapid loss of hair and gastrointestinal disorders. The sulfate (Tl_2SO_4) has been used as a rodenticide and insecticide.

The flowers of hydrangeas may be pink or blue; blue ones (above) are so colored because they contain an aluminum anthocyanin compound. In some cases, pink hydrangea flowers can be made to turn blue by the addition of alum to the soil.

Gallium has few uses, one of the main ones being in electronic components such as light-emitting diodes (left), which utilize gallium arsenide or phosphide.

Fact entries

Boron was isolated in June 1808, almost simultaneously by the French chemists Joseph Gay-Lussac (1778—1850) and Louis-Jacques Thénard (1777—1857), and only nine days earlier than the Englishman Humphry Davy (1778—1827). It is named after borax, its main compound. At. no. 5; at. mass 10.810; m.p. about 2,200°C; b.p. about 2,550°C.

Aluminum was first isolated (in crude form) in 1825 by the Danish scientist Hans Oersted (1777—1851). Its name is derived from the Latin *alumen,* meaning alum. At. no. 13; at. mass 26.982; m.p. 660°C; b.p. 2,327°C.

Gallium was discovered spectroscopically in 1874 and isolated in the following year by the French chemist Paul-Émile Lecoq de Boisbaudran (1838—1912), who named it after the Latin word *(Gallium)* for his homeland. At. no. 31; at. mass 69.720; m.p. 29.8°C; b.p. about 2,400°C.

Indium was discovered in 1863 by the German scientists Ferdinand Reich (1799—1882) and Theodor Richter (1824—1898). Its name derives from the Latin *indicum,* meaning indigo, the color of the element's dominant spectral lines. At. no. 49; at. mass 114.820; m.p. 155°C; b.p. about 2,000°C.

Thallium was first identified spectroscopically in 1861 by the British scientist William Crookes (1832—1919). Characterized by a bright green line in its spectrum, the element's name comes from the Greek *thallos,* meaning green twig. At. no. 81; at. mass 204.370; m.p. 303.5°C; b.p. 1,457°C.

IIIA	IVA	VA
5 **B** 10.810	6 **PB** 207.200	7 **N** 14.007
13 **Al** 26.982	14 **Si** 28.086	15 **P** 30.974
31 **Ga** 69.720	32 **Ge** 72.590	33 **As** 74.922
49 **In** 114.820	50 **Sn** 118.690	51 **Sb** 121.750
81 **Tl** 204.370	82 **C** 12.011	83 **Bi** 208.980

Carbon is the first member of Group IVA of the Periodic Table. Apart from its uniqueness in the number of compounds it can form, carbon resembles silicon in many respects; the degree of similarity with other Group IVA elements decreases down the group.

Inorganic carbon

Carbon (C) is unique among the chemical elements in the number and variety of the compounds it can form, many of which are the basis of living matter and the subject of organic chemistry. This article mainly considers some of the other compounds of carbon, so-called "inorganic" carbon.

The amorphous form of the element — charcoal and lamp black, for example — is the most common, but there are also two well-known crystalline forms: diamond, one of the hardest of known solids, and graphite, one of the softest. In diamond, each carbon atom has four other similar atoms at the points of a tetrahedron. Diamonds occur naturally in the Earth's crust, although there are few economic sources, principally in central and southern Africa and the Soviet Union. The conversion of carbon to diamond was first claimed in 1880. After many other false claims made over the last hundred years, graphite was converted to diamond by General Electric (USA) in 1955, using very high pressure (70,000 atmospheres) and temperature (3,600°F, or 2,000°C) for several months, with liquid iron as a solvent and catalyst (other transition metals are also used). Synthetic diamonds are now made industrially, for use as abrasives.

Graphite, which occurs naturally in many locations but is also made synthetically, has a crystal structure composed of parallel layers of hexagonal rings of carbon atoms. It has many applications: in dry batteries, lubricants, paint, and "lead" pencils.

Amorphous carbon occurs as natural organic matter, converted with time into coal, oil or gases. Various forms of carbon are obtained by thermal treatment — the hard coal anthracite is calcined (heated to a relatively high temperature for a prolonged period) to remove noncarbon constituents. Residual oils from petroleum refining are solidified to petroleum coke (for electrodes for aluminum refining) by heating. Carbon black, obtained by the vapor-phase decomposition of oils, with partial combustion, is used in motor tires and black printing inks. Charcoal is obtained by the destructive distillation of wood, sugar or other carbon-containing materials. "Active" charcoals are produced by selective oxidation to give products with very large surface areas, used as absorbents and catalysts.

Most carbon of fossil origin is burnt as fuel, giving carbon dioxide, CO_2. During the last century, the amount of CO_2 in the atmosphere has increased by about 15 per cent (to about 335 parts per million). It has been suggested that this carbon dioxide traps heat which would otherwise escape into space — the "greenhouse effect" — causing a predicted increase in the global average temperature of about 2 per cent by the middle of the next century. Opinions differ about the possible effects of this climate change; it will undoubtedly alter plant growth patterns, although many crops respond favorably (in greenhouses) to higher CO_2 levels.

Carbon dating

Carbon in living matter consists of a mixture of three isotopes more-or-less in equilibrium. They are stable ^{12}C, 98.9 per cent; stable ^{13}C, 1.1 per cent; and radioactive ^{14}C, 1 in 10^{12} parts. Exchange of carbon with the atmosphere ceases when an organism dies, so the carbon-14, with a half-life of 5,730 years, then decays, reducing to only 1 part in 10^{16} over about 70,000 years. Measurement of the proportion of carbon-14 in the total carbon of

Structure of diamond

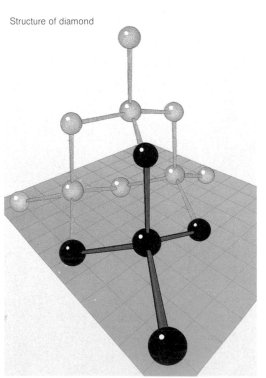

Carbon has two crystalline forms: diamond (above and left), with each atom bonded to four others in a tetrahedral arrangement; and graphite (below and right), with six-membered rings.

Structure of graphite

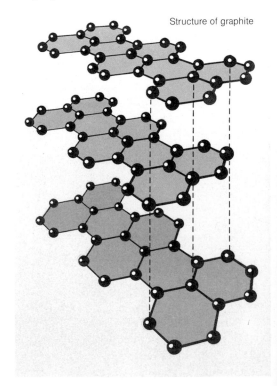

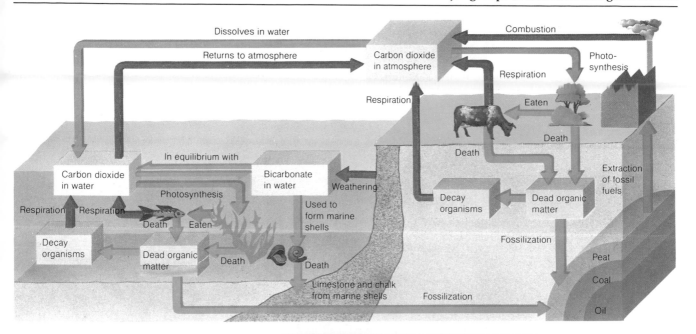

an archeological specimen of material that has once lived therefore gives an estimate of the age of the sample. Used with wood, the method can measure dates up to about 10,000 years ago. A limitation so far has been the size of sample required for the test, but recent improvements mean that as little as a thirty-thousandth of an ounce (a thousandth of a gram) can be tested — a single thread from a garment, for example.

In carbon-14 testing the wood of American bristlecone pine trees (the oldest living organisms), individual tree growth rings can be dated up to about 8,200 years ago with an accuracy of better than 5 per cent. Much of the error comes from contamination of samples with other carbon-containing matter, such as modern rootlets. There have also been some variations from the expected level of carbon-14 radioactivity since 1900, caused by ^{14}C-free carbon dioxide in the atmosphere from burned fossil fuels, and by radioactivity from atomic explosions.

Manmade fuels

The conversion of carbon to "synthesis gas," a mixture of carbon monoxide (CO) and hydrogen, could form the foundation of a chemical industry based on coal when the cost and scarcity of crude oil makes the present petrochemical methods uneconomic. Mixtures of carbon monoxide and hydrogen, in various proportions, can be passed

over different catalysts at varying high pressures and temperatures to produce a series of hydrocarbons, which can be converted to chemicals and fuels. Methanol can be produced in a similar way:

$$CO + 2H_2 \rightarrow CH_3OH$$

but can now also be obtained by improved low-pressure processes. The only commercial plant for making synthetic liquid fuels is in South Africa, which has inexpensive coal (due to low labor costs) and where a source of liquid fuel independent of crude oil is required for political reasons.

The carbon cycle is a complex chain of processes in which carbon compounds circulate among the air, water, living organisms and minerals. The illustration above shows the principal stages.

Carbon fibers are utilized extensively as reinforcing agents for lightweight high-strength plastics, used to make articles such as tennis rackets (as in the photograph, left) and other sports equipment.

Fact entries

Carbon has been known since prehistoric times. It occurs free in the Earth's crust as graphite, diamond and amorphous forms such as coal. Its name is derived from the Latin for charcoal, *carbo*. At. no. 6; at. mass 12.011; m.p. about 3,550°C; b.p. about 4,200°C.

Allotropy, the existence of an element in two or more distinct physical forms, is exemplified by carbon, which exists naturally in three main forms: graphite, diamond and amorphous carbon. Other elements that exhibit allotropy include oxygen, phospho-

rus, sulfur and tin. Allotropes may be monotropic or enantiotropic. In monotropy, one allotrope is the most stable under all conditions. Carbon, oxygen and phosphorus are monotropic: graphite is the stablest form of carbon; diatomic oxygen (O_2) is more

stable than the triatomic form (ozone, O_3); and red phosphorus is stabler than is white. In enantiotropy, different allotropes are stable under different conditions; sulfur and tin are enantiotropic. Sulfur forms rhombic and monoclinic crystals, the former being

the most stable form below about 95°C and the latter the most stable between 95°C and 120°C. Similarly, the gray form of tin is the most stable below about 13°C and the white allotrope is the stablest form at higher temperatures.

IIIA	IVA	VA
5 **B** 10.810	6 **C** 12.011	7 **N** 14.007
13 **Al** 26.982	14 **Si** 28.086	15 **P** 30.974
31 **Ga** 69.720	32 **Ge** 72.590	33 **As** 74.922
49 **In** 114.820	50 **Sn** 118.690	51 **Sb** 121.750
81 **Tl** 204.370	82 **Pb** 207.200	83 **Bi** 208.980

The elements silicon to lead comprise most of Group IVA of the Periodic Table. All have practical applications, although silicon (used in glass) and tin (used in canning) are the most commonly encountered.

Tin is widely used for coating steel cans — a use for which it is particularly suitable because it does not corrode easily, is not toxic and adheres firmly to the underlying steel.

Silicon to lead

Silicon and the elements below it make up most of Group IVA of the Periodic Table (the other element in this Group, carbon, is dealt with in the previous article and in the section on organic chemistry). They vary from the nonmetallic silicon (Si), through semimetallic germanium (Ge), to the metals tin (Sn) and lead (Pb). The first two elements have recently found use in semiconductors, whereas the other two have long been employed as the principal ingredients in a range of low-melting alloys.

Silicon and germanium

Silicon is the second most abundant element, making up to 25 per cent of the Earth's crust. It occurs in various forms combined with oxygen in the many types of silicates (more than 800 natural silicates are known) with the tetrahedral SiO_4 unit as the primary structure, and as silica (SiO_2). Silica occurs naturally as the three crystalline forms quartz, tridymite and cristobalite (a fourth form was produced in the first atomic bomb explosion in New Mexico).

Quartz occurs in granite and sandstone and is used for its piezoelectric and optical properties. Ordinary glass contains about 75 per cent silica, borosilicate glass has more than 80 per cent (with 12 per cent B_2O_3), and lead glass (for lenses and ornaments) contains about 55 per cent silica with 33 per cent lead oxide. The basic structural geometry of most forms of silica is a tetrahedral arrangement of a silicon atom surrounded by four oxygen atoms. From this, an infinite number of arrangements can be constructed.

Metallurgical grades of silicon are obtained commercially by reducing silica with carbon in an electric furnace. For semiconductor use, crude silicon is converted to a halide (usually $SiHCl_3$), which is reduced with pure hydrogen. The silicon is further purified by crystallization from the melt. Other techniques of purification, including zone refining and vapor deposition, are also used to supply suitable silicon for making transistors. In 1950, 2 ounces(50 grams) of high purity silicon produced 50 transistors. In 1980, the same amount provided a million, as integrated circuits.

Heating silicon and carbon in an electric furnace at 3,600—4,700°F (2,000—2,600°C) yields silicon carbide (SiC, carborundum), an important abrasive.

Germanium is comparatively rare, but occurs in the flue dust of some coals, and in copper and zinc ores in small but useful amounts. It has been widely used, in very pure form, for semiconductors, although silicon has become more important recently in these applications. The main uses of germanium now lie in the field of infrared optics and, possibly, solar cells. Even so, the total world use is less than 110 tons (100 metric tons) per year. The chemical reactivity of elemental germanium lies between silicon and tin, and it forms compounds in oxidation states II and IV.

Tin and lead

Tin was one of the earliest metals known and its use has closely followed the growth of civilization — food canning and preservation developed with tin-coated steel; fast machinery depended on tin-based bearings (Babbitt metal); printing expanded with tin-containing type metal; many communications devices first depended on tin-based solder; and tin alloys are used for musical instruments and (especially) organ pipes. But the metal is comparatively rare in nature (only 0.004 per cent of the Earth's crust), and is now sufficiently expensive for tin cans to be collected and the tin extracted and recycled.

Metallic tin has two definite crystal structures — tetragonal and cubic — with a transition temperature of about 55°F (13°C). When bent, tin makes a cracking "cry" as its crystals crush together. Tin metal is nontoxic, and is not corroded by weak acids and alkalis, which makes it suitable for containing foods (as the coating for steel cans). Oxygen or dry air forms a protective oxide coating on tin, which thickens with increasing temperature. Baths of molten tin are the basis of the float glass process for making plate glass.

Tin forms many tin (II) (Sn^{2+}) and tin (IV) (Sn^{4+}) compounds, such as the halides. Tin (II) chloride ($SnCl_2$) is used in tin-plating solutions, and tin (IV) chloride ($SnCl_4$) is an intermediate for making organo-tin compounds, employed in stabilizing plastics and as biocides. Tin compounds are also used in dyeing silk to "weight" the fabric and improve its texture.

Lead — like tin, known since antiquity — is widely distributed in the Earth's crust (but less common than zinc or uranium), usually as lead sulfide (PbS, galena). The metal is readily obtained by conventional smelting. It is very soft

The eroded sandstone pinnacles of Bryce Canyon, in the United States (left), and the stained glass window (below) both consist largely of silica (SiO_2) in the form of quartz in the sandstone, and mixed with lead oxide in the stained glass. (Most old stained glass also contains various metal compounds to make it colored — gold, for instance, which produces red glass.)

and malleable, and is easily cast, joined and converted into pipes or sheets. The principal modern use of the metal is in the electrodes of storage batteries (for motor vehicles). Other applications include alloys for type metal, bearings and solders. It is the densest common metal (11.34 grams per cubic centimeter), and so is used as ballast weight, and for shielding X-ray equipment and to absorb gamma rays.

Chemically, lead forms compounds in oxidation states II or IV. Its oxides include litharge (PbO), used in lead glazes on ceramics, and red lead (Pb_3O_4), the pigment of corrosion-resistant lead paints. Basic lead carbonate, ($PbCO_3$)$_2$. $Pb(OH)_2$, is the active component of "white lead" paints, and lead chromate ($PbCrO_4$) is the pigment chrome yellow. Most lead salts are very insoluble in water — the main exceptions are the nitrate and the acetate. Impure crystals of galena (PbS) were used in early radio "crystal sets" as a natural (and very crude) point-contact diode, long before the use of synthetic germanium and silicon devices. Tetraethyl lead, $Pb(C_2H_5)_4$, is added to gasoline to improve its antiknock rating — it is made by reaction of sodium-lead alloy with ethyl chloride. Because of the toxicity of lead, however, medical and environmental experts are applying pressure on governments and industries to reduce the use of these compounds, despite their technical efficacy. Lead compounds are absorbed in the body — about 10 per cent of ingested lead reaches the bloodstream. More than 90 per cent of such lead is eventually deposited in the bones, where it appears to be inert, but excess lead in the tissues can cause severe colic and adverse nervous effects. The largest source of lead in the environment is present naturally in soil, averaging about 16 parts per million.

Galena, the silver-gray crystals in the photograph (left), is the principal ore of lead. It is one of the most widely distributed sulfide ores and occurs in many different types of rocks. In addition to lead, some galena deposits contain silver and are therefore mined as a source of this metal also.

Fact entries

Silicon was first isolated in 1824 by the Swedish chemist Jöns Berzelius (1779—1848). Its name is derived from the Latin *silex*, meaning flint. At. no. 14; at. mass 28.086; m.p. 1,410°C; b.p. 2,355°C.

Germanium was predicted by the Russian chemist Dmitri Mendeleyev (1834—1907) in 1871 but was not discovered until 1886, by the German chemist Clemens Winkler (1838—1894), who named it after his homeland. At. no. 32; at. mass 72.590; m.p. 937°C; b.p. about 2,700°C.

Tin, in the form of bronze (an alloy with copper), has been known since at least 3500BC. Its chemical symbol, Sn, comes from the Latin word for the element, *stannum*. It occurs as three allotropes, one (gray tin) a powdery form at low temperatures. At. no. 50; at. mass 118.690; m.p. 231.9°C; b.p. 2,507°C.

Lead has been known as a metal since at least 4000BC. Its chemical symbol, Pb, is derived from the Latin word for the element, *plumbum*. At. no. 82; at. mass 207.200; m.p. 327.4°C; b.p. 1,740°C.

	IIIB	IVB	VB	VIB	VIIB		VIII		IB
	21 Sc 44.956	22 Ti 47.900	23 V 50.941	24 Cr 51.996	25 Mn 54.938	26 Fe 55.847	27 Co 58.933	28 Ni 58.700	29 Cu 63.546
	39 Y 88.906	40 Zr 91.220	41 Nb 92.906	42 Mo 95.940	43 Tc (97)	44 Ru 101.070	45 Rh 102.906	46 Pd 106.400	47 Ag 107.868
	57 La 138.906	72 Hf 178.490	73 Ta 180.948	74 W 183.850	75 Re 186.207	76 Os 190.200	77 Ir 192.220	78 Pt 195.090	79 Au 196.967
	89 Ac (227)	104 (261)	105 (262)						

The transition metals

The 21 elements that form Groups IVB (excluding Element 104), VB (excluding Element 105), VIB, VIIB and VIII of the Periodic Table are called the transition metals. They can be considered as three series: titanium to nickel, zirconium to palladium, and hafnium to platinum.

Chemically, they have many similarities, although each has individual features. In each element of the series, the filling of the outer shell of 8 electrons is interrupted (in the horizontal period) to bring the penultimate shell from 8 to 18 electrons. Those of the first series have several oxidation states, the oxides becoming more acidic with increasing oxidation number. Many oxidation state II and III compounds and related complexes are formed, but high oxidation states occur only with chromium (V and VI), manganese (V, VII and VIII) and iron (V and VI).

The first transition series

These metals — titanium (Ti), vanadium (V), chromium (Cr), manganese (Mn), iron (Fe),

cobalt (Co) and nickel (Ni) — are all of major industrial importance. Iron is by far the most commonly used structural metal, usually as an alloy with other metals of the series. Of the many types, probably the most familiar are mild steel and the stainless steels; the latter contain 16 to 26 per cent chromium and 6 to 22 per cent nickel (typically, 18:8 Cr:Ni). Stainless steels have good corrosion resistance and strength at high temperatures, and can be fabricated relatively easily. The last three elements of the first series — iron, cobalt and nickel — show spontaneous magnetism. Iron magnets in compasses made practical the science of navigation, and modern magnetic oxides of transition metals are essential components of recording tape.

Titanium minerals are widely distributed (it is the ninth most common element), mainly as the oxide rutile (TiO_2) and as the iron compound ilmenite ($TiFeO_3$). Titanium metal is obtained in an energy-intensive process, by reducing the chloride with magnesium metal (the Kroll process). It is used to make components that have to retain their strength at high temperatures — aerospace components, for example.

Pure titanium dioxide is a nontoxic white pigment with high opacity used in white paint (for its hiding power) and papermaking. It is made either by dissolving ilmenite in sulfuric acid to remove the iron impurities, followed by precipitation of the hydroxide which is calcined, or (in a process that is more difficult but produces less effluent) by conversion of rutile to the chloride which is burnt in oxygen to give the oxide. Titanium tetrachloride ($TiCl_4$) is a corrosive liquid which reacts with moisture to produce white smoke, which is used for smokescreens and sky-writing. Organic titanium compounds are used as catalysts, especially in the production of the plastic polypropylene.

Vanadium compounds are widely distributed in small amounts in nature; most are obtained as by-products from the processing of other ores, such as those of uranium and phosphorus. Pure vanadium compounds have several oxidation states, and many are reddish in color. Like other transition elements, vanadium and its compounds — particularly vanadium pentoxide (V_2O_5) — are used as catalysts in many petrochemical processes, in ceramic glazes, and are proposed for catalytic reduction of car-exhaust pollution. Vanadium metal is used, with iron and other transition metals, in high-performance alloys, especially tool steels. "Ferrovanadium" (50 to 70 per cent vanadium, with iron) is made by reducing fused vanadium oxides and iron with aluminum metal.

Vanadium compounds are highly toxic to animals, disrupting the metabolism of sulfur and cholesterol. Vanadium-containing ash from fuels can affect the respiratory organs.

Chromium, like the other elements of this group (apart from titanium), forms many highly colored salts. The element itself is found in nature mainly as the mineral chromite, with an

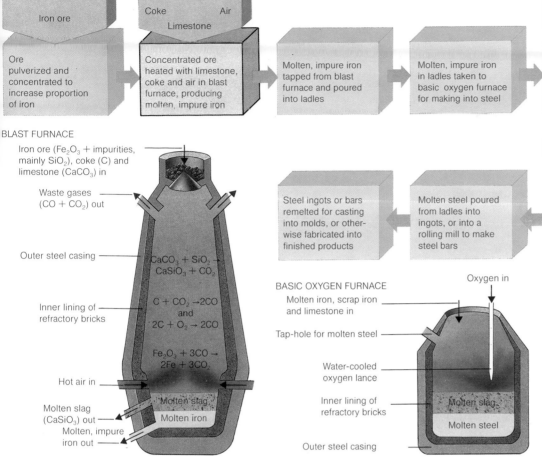

BLAST FURNACE

Iron ore (Fe_2O_3 + impurities, mainly SiO_2), coke (C) and limestone ($CaCO_3$) in

Waste gases (CO + CO_2) out

Outer steel casing

Inner lining of refractory bricks

$CaCO_3 + SiO_2 \rightarrow$
$CaSiO_3 + CO_2$

$C + CO_2 \rightarrow 2CO$
and
$2C + O_2 \rightarrow 2CO$

$Fe_2O_3 + 3CO \rightarrow$
$2Fe + 3CO_2$

Hot air in

Molten slag ($CaSiO_3$) out
Molten, impure iron out

Molten slag
Molten iron

BASIC OXYGEN FURNACE

Molten iron, scrap iron and limestone in

Tap-hole for molten steel

Oxygen in

Water-cooled oxygen lance

Inner lining of refractory bricks

Molten slag
Molten steel

Outer steel casing

Flow diagram boxes:

Iron ore → Ore pulverized and concentrated to increase proportion of iron

Coke, Air, Limestone → Concentrated ore heated with limestone, coke and air in blast furnace, producing molten, impure iron

Molten, impure iron tapped from blast furnace and poured into ladles

Molten, impure iron in ladles taken to basic oxygen furnace for making into steel

Oxygen, Scrap iron, Limestone → Molten iron, scrap iron limestone and oxygen heated in oxygen furnace, producing refined, basic steel

Alloying metals → Molten steel poured into ladles and alloying metals added (if desired) to produce specific alloy steels

Molten steel poured from ladles into ingots, or into a rolling mill to make steel bars

Steel ingots or bars remelted for casting into molds, or otherwise fabricated into finished products

The main stages in the commercial production of iron and its chief alloy, steel, are illustrated in the flow diagram (above). The key stages are the reduction of the iron ore to impure iron in a blast furnace (the details of which are shown in the diagram far left), and the refining of this impure iron in an oxygen furnace (left) to produce basic steel. The iron from the blast furnace contains up to 5 per cent of impurities, which make the metal brittle. These are removed by oxidation at a high temperature in the oxygen furnace — a process which is so efficient that scrap iron can be added for direct conversion to steel. The basic steel produced consists of iron with 0.1 to 0.3 per cent carbon; other metals may then be added to make alloy steels.

average composition of 68 per cent Cr_2O_3 and 32 per cent FeO (in practice the composition varies to include smaller amounts of the oxides of aluminum, magnesium and silicon). Chromite occurs in South Africa, the Soviet Union, the Philippines and Zimbabwe. The metal is obtained by reduction of the ore in an electric furnace, using either aluminum or silicon:

$$Cr_2O_3 + 2Al \rightarrow 2Cr + Al_2O_3$$
$$\text{or } 2Cr_2O_3 + 3Si \rightarrow 4Cr + 3SiO_2$$

Reduction with carbon yields ferrochromium, used in producing chromium alloys for high-strength steels. Chromium oxides, apart from their chemical uses in leather tanning and pigment manufacture, are mixed with magnesium and alu-

minum oxides for use in refractory bricks for lining steelmaking furnaces.

Manganese, like chromium, is mainly employed in special steels. The oxide (MnO_2), which occurs naturally as pyrolusite, is also made synthetically for use in batteries. Cobalt and nickel are also used in special steels and ceramics. Workable ore deposits are relatively rare. The ore is reduced to impure metal for alloy production with iron.

The human body contains about a fifth of an ounce (about 5 grams) of iron, of which two-thirds is in hemoglobin, the red pigment in the blood. Red meat, eggs, carrots, fruit and green vegetables provide the iron required daily in the diet.

Fact entries

Titanium was first isolated in 1910 by the New Zealand metallurgist Matthew Hunter, although the element's compounds had been known much earlier. Its name is derived from the Titans in Greek mythology. At. no. 22; at. mass 47.900; m.p. 1,677°C; b.p. 3,277°C.

Vanadium was first discovered in 1801 by the Spanish mineralogist Andrés del Río and then rediscovered in 1830 by the Swedish chemist Nils Sefström (1787—1845), who named it after Vanadis, the Scandinavian goddess of beauty. At. no. 23; at. mass 50.941; m.p. 1,917°C; b.p. about 3,000°C.

Chromium was discovered in 1797 by the French chemist Louis-Nicolas Vauquelin (1763—1829). Its name comes from the Greek *chromos,* meaning color, because many of its compounds are highly colored. At. no. 24; at. mass 51.996; m.p. 1,900°C; b.p. 2,642°C.

Manganese was recognized as an element in 1774 by the Swedish chemist Karl Scheele (1744—1786) and isolated in the same year by the Swedish mineralogist Johan Gahn (1745—1818). Its name is probably a corruption of Magnesia, the name of a region of ancient Greece. At. no. 25; at. mass 54.938; m.p. 1,244°C; b.p. 2,095°C.

Iron has been known since at least 3000BC, although it was not in wide use until about 1200BC. Its chemical symbol, Fe, derives from the Latin name for the element, *ferrum.* At. no. 26; at. mass 55.847; m.p. 1,535°C; b.p. 3,000°C.

Cobalt was isolated in about 1735 by the Swedish chemist Georg Brandt (1694—1768), although its blue-colored compounds had been known centuries earlier. Its name derives from the German *kobold,* meaning goblin. At. no. 27; at. mass 58.933; m.p. 1,493°C; b.p. about 3,100°C.

Nickel was isolated in 1751 from an ore containing nickel arsenide by the Swedish chemist Axel Cronstedt (1722—1765). The ore he used was once called (in German) *Kupfernickel,* from which the element later derived its name. At. no. 28; at. mass 58.700; m.p. 1,555°C; b.p. 2,837°C.

IIIB	IVB	VB	VIB	VIIB	VIII			IB
21 Sc 44.956	22 Ti 47.900	23 V 50.941	24 Cr 51.996	25 Mn 54.938	26 Fe 55.847	27 Co 58.933	28 Ni 58.700	29 Cu 63.546
39 Y 88.906	40 Zr 91.220	41 Nb 92.906	42 Mo 95.940	43 Tc (97)	44 Ru 101.070	45 Rh 102.906	46 Pd 106.400	47 Ag 107.868
57 La 138.906	72 Hf 178.490	73 Ta 180.948	74 W 183.850	75 Re 186.207	76 Os 190.200	77 Ir 192.220	78 Pt 195.090	79 Au 196.967
89 Ac (227)	104 (261)	105 (262)						

The block of transition metals from zirconium to technetium and from hafnium to rhenium exhibit similar properties and can be treated as a single group. All are rare in nature — apart from technetium, which does not occur naturally at all — and they tend to have only a few, highly specialized uses. Zirconium (found in the gemstone zircon), tungsten and molybdenum (both used in various alloys) are probably the most familiar.

The eight elements that form the first halves of the second and third transition series can be treated as one group. They are zirconium (Zr), niobium (Nb), molybdenum (Mo), technetium (Tc), hafnium (Hf), tantalum (Ta), tungsten (W) and rhenium (Re). They vary in abundance from the relatively rare to complete nonoccurrence in nature. All are metals, with considerable similarities in their properties, notably in their vertical relationships in the Periodic Table. Zirconium and hafnium, for example, have generally similar metallic properties, oxidation states and compounds, as do niobium and tantalum, and — especially — molybdenum and tungsten. All the elements have several oxidation states, and they form many complex compounds.

Zirconium occurs widely, mainly as the mineral zircon ($ZrSiO_4$); hafnium is usually present as an impurity (about 2 per cent). One of the most important uses of zirconium is as a structural material in atomic reactors and for cladding the uranium fuel elements, because of its extremely low neutron absorption. For this purpose, it is usually separated from hafnium — which readily absorbs neutrons — by liquid-liquid countercurrent extraction. The metals are obtained by a process similar to the Kroll process using molten magnesium.

Both zirconium and hafnium have high melting points — 3,374°F (1,857°C) and 4,040°F (2,227°C), respectively. The powdered metal is used in smokeless flash powders, blasting caps and pyrotechnics. Since the 1950s hafnium has been employed as a nuclear control-rod material, especially in the relatively compact reactors used in submarines. Small amounts of hafnium may be added to high-strength nickel-base alloys used for cutting tools and aircraft engine parts. Zirconium steels are also widely used in high-strength applications, such as rock drills. Some of the best superconducting metals are niobium-zirconium alloys, containing 20 to 40 per cent zirconium.

Niobium and tantalum have similar separation problems to those of zirconium and hafnium, and for many years there was confusion about the separate identity of niobium and tantalum. The two metals occur together, usually as a mixed oxide; tantalum is also obtained as a residue from tin production. Both sources involve solvent extraction with hydrogen fluoride, followed by liquid-liquid separation in methyl isobutyl ketone. In this way, the fluorides of tantalum, niobium, titanium, zirconium, iron and manganese can be separated.

Niobium, as "ferroniobium," is used in making high-strength steels — especially those for oil and gas pipelines which operate below —13°F (—25°C). It has a low absorption for thermal neutrons and is used with zirconium to increase alloy strength for reactor cladding. Most superconducting devices employ niobium-titanium alloys, because they can be easily fabricated into wire. The carbides of niobium and tantalum are added to hard metals to improve toughness, for use in drill bits and turbine blades. Tantalum was employed as the filament in early electric lamps (replaced by tungsten after about 1910) and today it is used extensively as the metallic part of small capacitors in electronic devices.

Molybdenum is used as an alloying agent in armorplating (which basically consists of alloy steel) because it is one of the most effective elements for increasing the hardenability and toughness of steel. In addition, molybdenum is highly corrosion resistant, and is unusual in retaining its strength and hardness at extremely high temperatures — properties that make the metal itself and its alloys very useful for industrial chemical equipment.

Technetium resembles rhenium chemically although, unlike rhenium, it does not occur naturally and all its isotopes are unstable and radioactive. Produced artificially in nuclear reactors, technetium isotopes with mass numbers from 92 to 107 have been prepared; the most commonly available, ^{99}Tc, has a half-life of about 2.1 x 10^5 years and is used in radiotherapy of thyroid disorders. The other principal long-lived isotopes are ^{97}Tc (half-life 2.6 x 10^6 years) and ^{98}Tc (half-life 1.5 x 10^6 years).

Two natural isotopes of rhenium occur: ^{187}Re, the most common, and ^{185}Re. The former is radioactive, with a half-life of about 10^{11} years. Chemically, rhenium has seven oxidation states, of which the highest is the most stable. The metal is comparatively rare and expensive, and is mainly used for its high melting point in high-temperature thermocouples.

Molybdenum and tungsten are industrially the most important elements of this group. Both are components of many important alloys, especially armor-plate, tool steels, and high-strength steels for aircraft engines. Both metals also have many colored compounds, with several oxidation states. Tungsten gets its chemical symbol W from wolframite, the name of its principal ore. It has the highest melting point — 6,170°F (3,410°C) — of all metals. About three-quarters of the world's reserves are in China.

Tungsten carbide (WC) is made by heating the powdered metal with carbon in hydrogen at about 2,700°F (about 1,500°C). The carbide powder is pressed into shape (usually with cobalt) to provide hard cutting edges for saws and drills. Molybdenum sulfide (MoS_2) has a platelike crystal structure — similar to graphite — so the molecules slide against each other. This property is exploited in high-performance lubricants. Most of the world's molybdenum comes from one large deposit (containing about 0.4 per cent MoS_2) at Climax in Colorado, United States.

Zirconium, hafnium and niobium are used in nuclear reactors, such as the one shown (left). Zirconium and niobium, or sometimes an alloy of the two, are used to clad the fuel elements because they allow neutrons to pass through them easily. In contrast, hafnium is used in some nuclear reactors for its high neutron absorption, a property that makes it valuable as a control-rod material.

Tungsten carbide (WC) is one of the most important compounds of tungsten because of its extreme hardness — making it useful for the cutting edges of machine tools (as below).

Fact entries

Zirconium was identified in 1789 by the German chemist Martin Klaproth (1743—1817) but was not isolated until 1824, by the Swedish chemist Jöns Berzelius (1779—1848). Its name derives from that of its main source, the semi-precious mineral zircon (zirconium silicate) At. no. 40; at. mass 91.220; m.p. 1,857°C; b.p. 3,577°C.

Niobium was first discovered in 1801 by the British chemist Charles Hatchett (1765—1847), who named it columbium (a name still used occasionally with the chemical symbol Cb). The element was rediscovered in 1844 by the German chemist Heinrich Rose (1795—1864), who distinguished it from tantalum and named it niobium after Niobe, the daughter of Tantalus in Greek mythology. At. no. 41; at. mass 92.906; m.p. 2,468°C; b.p. 4,927°C.

Molybdenum was identified in 1778 by the Swedish chemist Karl Scheele (1744—1786) and isolated in 1782 by another Swedish chemist, Peter Hjelm. Its name derives from the Greek molybdos, meaning lead, because it was once thought to be a lead ore. At. no. 42; at. mass 95.940; m.p. 2,622°C; b.p. about 4,825°C.

Technetium was the first element to be produced artificially, and its name comes from the Greek technetos, meaning artificial. It was discovered in 1937 by the Italian scientists Carlo Perrier and Emilio Segrè (1905—   ). At. no. 43; at. mass (of longest-lived isotope) 97; m.p. about 2,250°C; b.p. about 4,850°C; half-life of ^{97}Tc 2.6 x 10^6 years.

Hafnium derives its name from the Latin Hafnia for Copenhagen, Denmark, where the element was discovered spectroscopically in 1923 by the Dutch physicist Dirk Coster (1889—1950) and the Hungarian-Swedish chemist Georg von Hevesy (1885—1966). At. no. 72; at. mass 178.490; m.p. 2,227°C; b.p. 4,602°C.

Tantalum was discovered in 1802 by the Swedish chemist Anders Ekeberg (1767—1813), who named it after Tantalus in Greek mythology because its isolation had proved to be such a tantalizing task. At. no. 73; at. mass 180.948; m.p. 2,996°C; b.p. 5,429°C.

Tungsten derives its name from the Swedish words tung, meaning heavy, and sten, meaning stone. It was first isolated in 1783 by the Spanish scientists Fausto Elhuyar and his brother Juan, who obtained the metal from the mineral wolframite (iron and manganese tungstate) — hence the element's alternative name of wolfram and its chemical symbol W. At. no. 74; at. mass 183.850; m.p. 3,410°C; b.p. 5,900°C.

Rhenium was predicted in 1869 by the Russian chemist Dmitri Mendeleyev (1834—1907) but was not discovered until 1925, by the German chemists Walter Noddack (1896 –1960), his wife Ida (1896—   ) and Otto Berg, who named it after the River Rhine. At. no. 75; at. mass 186.207; m.p. 3,180°C; b.p. about 5,900°C.

IIIB	IVB	VB	VIB	VIIB		VIII		IB
21 Sc 44.956	22 Ti 47.900	23 V 50.941	24 Cr 51.996	25 Mn 54.938	26 Fe 55.847	27 Co 58.933	28 Ni 58.700	29 Cu 63.546
39 Y 88.906	40 Zr 91.220	41 Nb 92.906	42 Mo 95.940	43 Tc (97)	44 Ru 101.070	45 Rh 102.906	46 Pd 106.400	47 Ag 107.868
57 La 138.906	72 Hf 178.490	73 Ta 180.948	74 W 183.850	75 Re 186.207	76 Os 190.200	77 Ir 192.220	78 Pt 195.090	79 Au 196.967
89 Ac (227)	104 (261)	105 (262)						

The block of transition metals from ruthenium to palladium and from osmium to platinum are commonly called the platinum group metals. They usually occur together in nature, although they are rare and difficult to extract and separate from each other.

Six of the transition metals are known as the precious metals; they are ruthenium (Ru), rhodium (Rh), palladium (Pd), osmium (Os), iridium (Ir) and platinum (Pt). They occupy positions in the second and third triads of Group VIII. Although differing in atomic weight, the physical and chemical properties of each element are similar to those of the corresponding member of the other triad.

All the elements in this group are rare, but tend to occur together, and are obtained and separated only with difficulty. They are high-density metals with generally similar properties, and they resist oxidation, particularly in aqueous solution. They have high melting points, and can form extremely hard alloys (for electrical contacts and the tips of pen nibs). In finely divided form, such as platinum black, the metals are highly active catalysts.

Chemically, these elements have many different oxidation states — ruthenium, for example, has nine, from zero (in the carbonyl, $Ru(CO)_5$) to

eight in the stable tetroxide (RuO_4). There is a marked tendency to form stable complexes containing more than one metal atom — simple aqueous ions are uncommon, and virtually unknown with some of the elements. These properties can be explained by considering the valence electrons of the elements, which tend to be in the lower energy orbitals and so are more difficult to remove; this results in higher ionization energies, reflected in the strength of the metallic bonding. In the Periodic Table, the compounds of second row metals are less stable than the analogous ones of the third row metals, and the former therefore tend towards greater kinetic activity. The stability of the higher oxidation states decreases from left to right within each row.

Most of the platinum group metals are obtained from Canada, South Africa and the Soviet Union, often as minor components of copper ores. World production is less than 110 tons (100 metric tons) per year.

Modern refinery techniques depend on the ready solubility of platinum, palladium and gold in aqua regia (a 4:1 mixture of hydrochloric and nitric acids), and the ease with which gold can be reduced to metal from the chloride. Platinum is separated as ammonium chlorplatinate, which leaves palladium in solution. Material undissolved by aqua regia includes rhodium, ruthenium, osmium, iridium and silver. These metals are then separated by a series of fusion processes. More recently, a less energy-intensive solvent extraction method has been introduced.

Most platinum metal is used as a catalyst (usually as a rhodium-platinum gauze) in the conversion of ammonia to nitric oxide, for fertilizers

The most recent technique for extracting the platinum group metals (PGMs) is based on solvent extraction; part of a refinery in which this process is carried out is shown in the photograph (right). The details of the solvent extraction of PGMs are closely-guarded secrets; basically, however, a mixture of the metals in aqueous solution is mixed with a water-immiscible organic reagent that reacts with only one of the metals, thereby removing it from the aqueous phase and into the organic phase. The metal-reagent phase is then separated from the aqueous phase and processed to extract the metal in pure form. This process is repeated to obtain each of the metals in turn.

The seal in the photograph (left) is made of platinum, a precious metal used extensively in jewelry and other such ornamental objects because of its brilliant silver-white appearance and its malleability, as well as its resistance to corrosion by acids (it is, however, attacked by caustic alkalis).

The tips of pen nibs (right) are often made of osmium-iridium alloy, a substance that is very hard and has excellent corrosion resistance. Almost all osmium produced is used in alloys because the pure metal is brittle, even at high temperatures.

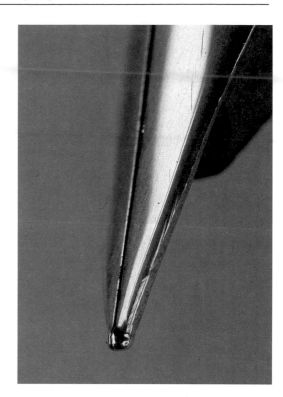

and explosives, and in the production of high-octane gasoline and aromatic compounds. Platinum- and palladium-containing reactors are installed in some cars, to detoxify exhaust gases. Oxidation catalysts, with mixed platinum-group metals, are used in air pollution control, to remove organic vapors and unpleasant odors. Other platinum alloys are used in electrical switching contacts. Platinum vessels (crucibles, for example) are used in laboratories because of their corrosion resistance; small amounts of rhodium or iridium are sometimes added to increase hardness.

The International Standard kilogram of mass is made from an alloy containing 90 per cent platinum and 10 per cent iridium, because the metal is so stable. Fuel cells, with electrodes made of platinum, are likely to become more common. The electrical resistivity of platinum is comparatively high, and the International Temperature Scale, between $-297°F$ and $1,166°F$ ($-183°C$ and $630°C$), is defined in terms of a resistance thermometer made from platinum wire.

Platinum forms many complex compounds, in which ammonia (NH_3), the chloride (Cl^-) ion or other groups (ligands) are bonded to a central platinum atom. Of the transition metals, platinum has the greatest tendency to bond to carbon. Some such compounds, notably "cisplatin" (cis-amminodichloroplatinum (II), $Pt(NH_3)_2Cl_2$), are used in cancer chemotherapy to treat tumors. The use of other platinum-group complexes in this field appears possible. Rhodium, iridium, and osmium were discovered in the early 1800s by working some native platinum ore from South America and separating the metals from aqua regia solutions of the impure platinum. Rhodium chemistry involves chiefly Rh(I) and Rh(III) compounds (with a few compounds up to Rh(VI)). All are readily decomposed by heat to give the powdered or sponge metal. Rhodium alloys are used in high-temperature thermocouples and pyrometers. Osmium is the metal of this group most readily attacked by air, forming the poisonous volatile osmium tetroxide (OsO_4), which has a characteristic odor.

Fact entries

Ruthenium was isolated in 1844 by the Russian chemist Karl Klaus (1796—1864). Its name comes from the medieval Latin *Ruthenia* for a region in central Europe (now part of the Soviet Union). At. no. 44; at. mass 101.070; m.p. about 2,450°C; b.p. about 4,150°C.

Rhodium was isolated in 1803 by the British chemist

William Wollaston (1766—1828), who named it from the Greek *rhodon,* meaning rose, because of the red color of many of the element's compounds in solution. At. no. 45; at. mass 102.906; m.p. 1,966°C; b.p. about 3,727°C.

Palladium was isolated in 1803 by William Wollaston and named after Pallas, a recently discovered aster-

oid. At. no. 46; at. mass 106.400; m.p. 1,555°C; b.p. 3,167°C.

Osmium was discovered in 1804 by the British chemist Smithson Tennant (1761—1815), who named it for the unpleasant smell of some of its compounds (the Greek *osme* means odor). At. no. 76; at. mass 190.200; m.p. about 2,700°C; b.p. about 5,500°C.

Iridium was discovered in 1804 by Smithson Tennant, who named it from the Latin *iris,* meaning rainbow, because of the variety of colors that are exhibited by its compounds. At. no. 77; at. mass 192.220; m.p. 2,450°C; b.p. about 4,500°C.

Platinum derives its name from the Spanish *platina,* meaning little silver,

because of its resemblance to the latter metal. It is not known when platinum was originally discovered but it was first reliably reported in South America in 1735 by the Spanish mathematician Antonio de Ulloa and was brought to Europe in 1741 by the British metallurgist Charles Wood. At. no. 78; at. mass 195.090; m.p. 1,773°C; b.p. 3,827°C.

IVA	VA	VIA
6 C 12.011	7 N 14.007	8 O 15.999
14 Si 28.086	15 P 30.974	16 S 32.060
32 Ge 72.590	33 As 74.922	34 Se 78.960
50 Sn 118.690	51 Sb 121.750	52 Te 127.600
82 Pb 207.200	83 Bi 208.980	84 Po (209)

Nitrogen is the first element in Group VA of the Periodic Table. It is by far the most abundant gaseous element on Earth, constituting more than three-quarters of the volume of the atmosphere.

The nitrogen cycle is a complex series of transformations in which atmospheric nitrogen is converted (by living organisms and lightning) into inorganic and organic nitrogen compounds and then converted back to gaseous nitrogen. The principal stages in the cycle are illustrated in the diagram below.

Nitrogen

Nitrogen (N), the first element in Group VA of the Periodic Table, is a colorless, odorless gas and an essential constituent of all living matter. It makes up approximately 78 per cent by volume of the Earth's atmosphere, and is found in plant and animal proteins.

Nitrogen is produced on a large scale by the fractional distillation of liquid air. This process involves the liquefaction of air and the removal of nitrogen at its boiling point, $-320.42°F$ ($-195.79°C$). Most nitrogen is used for the production of fertilizers from ammonia (NH_3) or nitric acid (HNO_3). The gas itself is, however, also used in the chemical, electrical and metals industries to provide an inert atmosphere, and in the food industry to prevent spoilage by mold. Liquid nitrogen is easily prepared and is used as a refrigerant.

There are few inorganic or mineral deposits containing nitrogen because most of its compounds are soluble in water. Deposits of sodium nitrate (Chile saltpeter) are found in Chile and some other areas with a dry climate, and these have been mined for use as a fertilizer or to make explosives.

The nitrogen cycle
On average, nitrogen makes up about 16 per cent of animal and vegetable proteins; the other constituents are carbon, hydrogen, oxygen and sulfur. Only a few simple organisms are able to use nitrogen directly from the air for the manufacture of the proteins they require for growth and tissue maintenance.

Plants in general take nitrogen from the soil in the form of nitrates ($-NO_3^-$), nitrites ($-NO_2^-$) and ammonium (NH_4^+) salts. Animals absorb most of the nitrogen they require from eating plants or other animals. There is therefore a vital relationship between plants, animals, the soil and the nitrogen in the air. This interdependence is known as the nitrogen cycle.

Nitrogen finds its way to the soil in rain water as dilute nitric acid (HNO_3) and nitrous acid (HNO_2) after the reaction between nitrogen and oxygen in the atmosphere initiated by lightning. Certain bacteria, found in the roots of plants such as beans, peas and clover, can convert nitrogen directly into proteins which plants then use. Nitric and nitrous acids react with bases in the soil to form salts, again which plants can utilize to form proteins. But not all this combined, or "fixed," nitrogen is held in the soil, and some is broken down back to nitrogen by bacteria and returns to the atmosphere.

Decaying plant tissue releases ammonia to the soil, which builds up there as ammonium salts. More bacteria convert these salts into nitrates and nitrites, which can be used once more by plants for the production of protein. Much of the nitrogen in animal protein is also returned to the soil when animals die and decay. Plant protein ingested by animals is excreted mainly as urea ($CO(NH_2)_2$) and this substance, returned to the soil, can react with water to give carbon dioxide and ammonia.

Nitrogen compounds
In the free state, nitrogen occurs as a triply-bonded diatomic molecule (N_2), which is very stable. In compounds it can combine with up to three other atoms or groups, forming covalent bonds and giving stable molecules. In the ammonium ion (NH_4^+), nitrogen shows a covalency of four; there is a lone pair of electrons, not normally associated with bonding, donated to the proton H^+, but all four N-H bonds in NH_4^+ are equivalent, just as are the C-H bonds in methane (CH_4).

Ammonia and hydrazine
Ammonia (NH_3) is the simplest hydride of nitrogen and one of the most important compounds of the element. A pungent-smelling and extremely soluble gas, it is produced in large

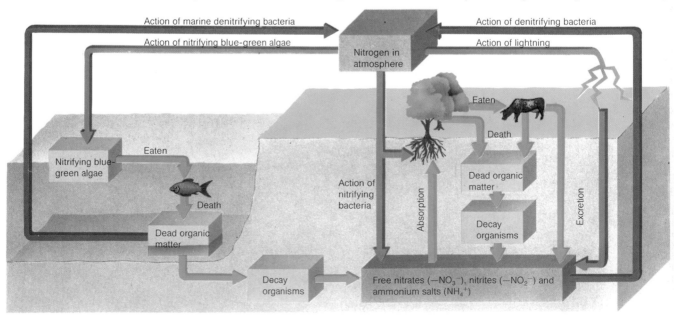

Action of marine denitrifying bacteria
Action of denitrifying bacteria
Action of nitrifying blue-green algae
Action of lightning
Nitrogen in atmosphere
Eaten
Death
Eaten
Nitrifying blue-green algae
Death
Action of nitrifying bacteria
Absorption
Dead organic matter
Excretion
Dead organic matter
Decay organisms
Dead organic matter
Decay organisms
Decay organisms
Free nitrates ($-NO_3^-$), nitrites ($-NO_2^-$) and ammonium salts (NH_4^+)

quantities for conversion to fertilizers and for use as a raw material in the production of nitric acid and nylon. It is also widely used as a refrigerant and a neutralizing agent.

Ammonia is produced commercially by the direct combination of nitrogen and hydrogen, reacted over a catalyst and at high pressure. Hydrogen for the process is derived mainly from natural gas; nitrogen is obtained by fractional distillation of liquid air.

A high temperature, high pressure, catalytic process for the production of ammonia was first developed by the German scientists Fritz Haber (1868—1934) and Carl Bosch (1874—1940) in the early part of this century. The Haber-Bosch process, using hydrogen derived from coal and air reacting over an iron catalyst, was used for many years to meet the increasing need for fertilizers.

Nitrogen also forms another hydride, hydrazine (N_2H_4). It is a colorless, unstable liquid and a strong reducing agent. Hydrazine was first used on a large scale as fuel for the German ME-163 rocket fighter, developed during World War II. Today it is used as a raw material for the production of pesticides, herbicides, pharmaceuticals, and foaming agents for certain plastics.

Oxides of nitrogen

Eight oxides of nitrogen are known, ranging from nitrous oxide (N_2O), a colorless unreactive gas — laughing gas — sometimes used as an anesthetic, to dinitrogen pentoxide (N_2O_5), which is unstable, and the compounds NO_3 and N_2O_6, which are also unstable and little studied. Nitrogen dioxide (NO_2) is a strong oxidizing agent and was once used in making sulfuric acid (to oxidize sulfur dioxide to sulfur trioxide).

Ammonia can be reacted with oxygen to form nitric oxide (NO), which is used industrially for the production of nitric acid (HNO_3). Nitric oxide is formed by passing air and ammonia over a platinum catalyst at 1,472—1,742°F (800—950°C), and then reacted with more oxygen to oxidize it further to nitrogen dioxide (NO_2). Nitrogen dioxide is dissolved in water, in a reaction which produces a lot of heat, to give nitric acid and more nitric oxide. Nitric acid of only about 60 per cent concentration is produced by this process, and special methods are needed to make acid of higher concentrations. Pure, 100 per cent nitric acid can be made by distilling a mixture of aqueous nitric acid and fuming sulfuric acid.

Nitric acid is a very powerful oxidizing agent, and gold, platinum, rhodium and iridium are the only metals not attacked by it. Some other metals though, such as aluminum, iron and copper, become "passive" in the acid due to the formation of a protective film of oxide on the metal. Most nitric acid is used for the manufacture of fertilizers, particularly ammonium nitrate (NH_4NO_3). Concentrated nitric acid is widely used for the production of nitrocellulose (guncotton) and glyceryl trinitrate (nitroglycerin), for making explosives.

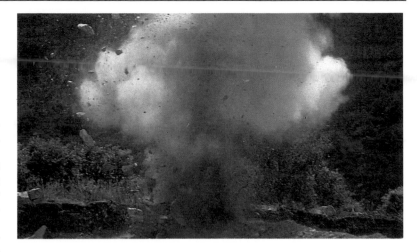

Dynamite (shown in use above) is a mixture of nitroglycerin (glyceryl trinitrate) and an inert substance such as kieselguhr or wood pulp, which is added to make the nitroglycerin safer to handle.

The nitrogen content of the soil is increased by spraying with nitrate fertilizers (above) and by the activity of nitrifying bacteria in the root nodules of certain plants (left), such as clover, beans and peas.

Fact entries

Nitrogen was first recognized by the French chemist Antoine Lavoisier (1734—1794), who named it azote, meaning "without life" because of its inability to support life. However, the element's discovery (1772) is usually credited to the British scientist Daniel Rutherford (1749—1819). Its present name, coined in 1790, derives from "niter" plus the suffix "—gen," meaning "niter forming," because of its presence in niter (potassium nitrate). At. no. 7; at. mass 14.007; m.p. —210.01°C; b.p. —195.79°C.

The Haber-Bosch process, developed in the early twentieth century, is still an important method of making ammonia — from which nitric acid and, in turn, explosives and fertilizers can be manufactured. The key reaction is the combination of one molecule of nitrogen (from air) and three molecules of hydrogen (now obtained from natural gas) to give two molecules of ammonia. This reaction is reversible, and significant amounts of ammonia can be obtained only by using high pressures (about 150 atmospheres), high temperatures (about 500°C) and a catalyst (iron plus promoters such as alumina).

IVA	VA	VIA
6 C 12.011	7 N 14.007	8 O 15.999
14 Si 28.086	15 P 30.974	16 S 32.060
32 Ge 72.590	33 As 74.922	34 Se 78.960
50 Sn 118.690	51 Sb 121.750	52 Te 127.600
82 Pb 207.200	83 Bi 208.980	84 Po (209)

The elements phosphorus to bismuth complete Group VA of the Periodic Table, which is headed by nitrogen. They exhibit increasing metallic properties down the group.

Powdered phosphate rock raises a cloud of dust which engulfs the handling machinery at a quarry in Jordan. Phosphate minerals — usually forms of calcium phosphate — are the major source of phosphorus and its compounds, particularly superphosphate for use as a fertilizer.

Phosphorus to bismuth

Phosphorus (P), arsenic (As), antimony (Sb) and bismuth (Bi) form a group of four elements with certain features in common. They occupy Group VA of the Periodic Table, with nitrogen at their head. Phosphorus is a highly reactive nonmetal, arsenic and antimony are poisonous metalloids, and bismuth is a true metal.

In some respects, phosphorus is similar to nitrogen. Both have typical nonmetallic characteristics and their oxides combine with alkalis to form salts; both are essential constituents of plant and animal tissue. But in other ways phosphorus is very different from nitrogen. It is a solid at normal temperatures and occurs in various structural arrangements, or allotropes, with a variety of colors and physical and chemical properties. One form of phosphorus is also very poisonous, as are many of its compounds, particularly its organic compounds.

The other elements in the group are unlike nitrogen, but similar to phosphorus. Arsenic and antimony exist as different allotropes, and certain of their compounds are toxic. From arsenic to bismuth the elements look more and more like metals, and some of their chemical characteristics confirm this. Because of the arrangement of the outer electrons in their atoms, each of the elements can form up to five chemical bonds with other elements or groups of elements. The pentavalent compounds, such as phosphorus pentachloride (PCl_5), and the trivalent compounds, such as arsine (AsH_3), are particularly important.

Phosphorus

Phosphorus can exist in the three main allotropic forms — white, red and black — each with very different chemical reactivities. White phosphorus is a soft waxlike solid, and when exposed to air it emits a faint green light which is visible in the dark. This glow, sometimes called phosphores-cence, gave phosphorus its name from the Greek word meaning light-bearer. It is caused by a complicated oxidation reaction and is an example of a general phenomenon known as chemiluminescence.

In an abundant supply of air, white phosphorus burns easily, giving off a highly toxic vapor. For this reason it is kept under water, with which it does not react. Because it burns so easily in the air, white phosphorus was once used to make matches. This was before it was known to be so poisonous, and many workers in the match industry suffered from a dreaded tooth and bone disease known as "phossy jaw."

Red phosphorus is a much more stable form of the element and is used today in the manufacture of matches, either on its own or as a sulfide (P_4S_3). It is formed by heating white phosphorus for several hours at 750°F (400°C) in the absence of air. It ignites in air only on strong heating.

Another form of the element — black phosphorus — is made after either heating white phosphorus at 430—700°F (220—370°C) for eight days with mercury as a catalyst, or by heating it under very high pressure for a shorter time.

Phosphorus occurs in both inorganic and organic forms in nature. The minerals phosphorite ($Ca_3(PO_4)_2$) and apatite ($3Ca_3(PO_4)_2.CaF_2$) are found as deposits around the world. The element is also present in the droppings of sea birds (called guano), found in large quantities on the coast of Peru and some Pacific islands. Guano has been mined to produce phosphorus fertilizers. The bones and teeth of animals contain the complex calcium phosphate salt hydroxyapatite, and every living cell in plants and animals contains phosphorus in some form or other — particularly the cells of brain and nervous tissue.

Phosphate-bearing rocks are mined on a large scale and treated with acid to produce a mixture of phosphorus and calcium salts, which can be used as fertilizers. These salts are also a raw material for making synthetic detergents and poultry and animal feeds. The use of phosphate salts for the production of synthetic detergents has declined recently because of the effects that these compounds have — discharged in effluent — in promoting plant growth in rivers and waterways.

Phosphorus, in the form of phosphate, is essential to the fertility of the soil and is continuously removed as plants are cropped. Unlike nitrogen, it is not replaced by a natural cycle and phosphorus-containing fertilizers need to be continually added. These can be prepared only from natural sources of phosphorus-bearing minerals. Many organic forms of phosphorus are among the most toxic substances known; some are used as insecticides, and others constitute deadly nerve gases.

Arsenic, antimony and bismuth

Arsenic, antimony and bismuth exist in bright

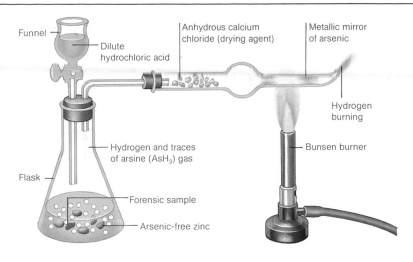

Funnel

Dilute
hydrochloric acid

Anhydrous calcium
chloride (drying agent)

Metallic mirror
of arsenic

Hydrogen
burning

Hydrogen and traces
of arsine (AsH₃) gas

Bunsen burner

Flask

Forensic sample

Arsenic-free zinc

Antimony compounds
were known to the ancient
Egyptians, who used the
sulfide (stibnite) as black
eye makeup, a feature
reflected in their sculpture
(far left).

In the Marsh test (left),
devised as a forensic test in
cases of suspected arsenic
poisoning, the gas arsine is
heated to form a metallic
mirror of arsenic.

metallic forms which are stable in air. A yellow form of arsenic can also be prepared by heating gray arsenic to 1,112°F (600°C) and collecting the vapor, but this is easily converted back to the gray form by gentle heating. Yellow antimony can be produced using a similar method, but this allotrope is stable only at low temperatures.

All three elements can be prepared from their sulfides, which occur naturally. Realgar (As₄S₄) and orpiment (As₄S₆) were known in early times as pigments and are mentioned by Pliny and Aristotle. Stibnite, the naturally occurring sulfide of antimony, was once used as a cosmetic and a medicine.

The elements are often used as alloying materials to improve the properties of metals in certain applications. Arsenic, for example, may be added to improve the high-temperature mechanical characteristics of alloys used in bearings. Bismuth forms alloys of extremely low melting point, and these are used as safety plugs on gas cylinders and in automatic sprinkler fire-extinguisher systems. Some compounds of the elements are very toxic, whereas others are not. The hydrides arsine (AsH₃) and stibine (SbH₃) are both very dangerous gases. Calcium arsenate and lead arsenate are used as pesticides, and sodium arsenite is a weed killer and a fungicide.

Ultrapure arsenic and antimony (more than 99.999 per cent pure) are now used in the electronics industry. Gallium arsenide (GaAs) is employed in semiconductor diodes, transistors and lasers, and the element itself is combined with selenium and other elements to make glasses which are transparent to infrared radiation. Indium arsenide (InAs) and gallium antimonide (GaSb) have been used as detectors of infrared.

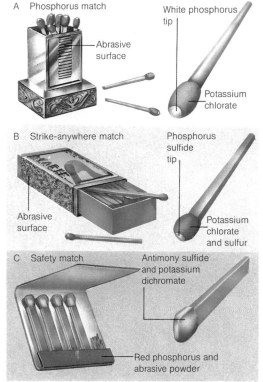

A Phosphorus match

Abrasive
surface

White phosphorus
tip

Potassium
chlorate

B Strike-anywhere match

Abrasive
surface

Phosphorus
sulfide
tip

Potassium
chlorate
and sulfur

C Safety match

Antimony sulfide
and potassium
dichromate

Red phosphorus and
abrasive powder

Phosphorus has long been
used in matches. The original phosphorus match (A)
— now illegal — had a
white phosphorus tip. It
was banned because of the
danger of phosphorus poisoning to match workers. A
strike-anywhere match (B)
uses phosphorus sulfide
instead. In safety matches
(C) the non-poisonous red
phosphorus is in the striking surface.

The two main allotropes of
phosphorus (below) have
different molecular structures. White phosphorus
(A) exists as tetrahedrally
joined groups of four
atoms, whereas red phosphorus (B) consists of
chains.

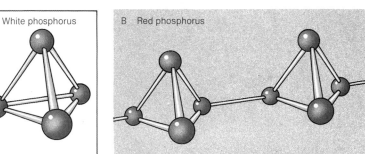

A White phosphorus

B Red phosphorus

Fact entries

Phosphorus was first isolated (from urine) by the German alchemist Hennig Brand in about 1669. In 1860 it was discovered independently by the English chemist Robert Boyle. It was named from the Greek *phosphoros*, meaning light-bearer, because the white allotrope glows in the dark. The red allotrope is not phosphorescent. At. no. 15; at. mass 30.974; m.p. (white) 44.1°C; b.p. (white) 280°C; m.p. (red) 600°C (under pessure).

Arsenic was identified by the German pharmacist Johann Schröder in 1649. The Latin word *arsenicum* means yellow orpiment (an arsenic trisulfide pigment). At. no. 33; at. mass 74.922; sublimes at 615°C without melting.

Antimony occurs mainly as its sulfide mineral stibnite (Latin *stibnum*), from which it derives its chemical symbol Sb. It was known to the Greeks and Romans. At. no. 51; at. mass 121.75; m.p. 630°C; b.p. approx. 1,350°C.

Bismuth has been known since about 1400 and was finally isolated by Caspar Newmann (1683—1737). Its name may derive from the Old German *vissmuth* (white matter). At. no. 83; at. mass 208.980; m.p. 271°C; b.p. approx. 1,470°C.

VA	VIA	VIIA
7 N 14.007	8 O 15.999	9 F 18.998
15 P 30.974	16 S 32.060	17 Cl 35.453
33 As 74.922	34 Se 78.960	35 Br 79.904
51 Sb 121.750	52 Te 127.600	53 I 126.904
83 Bi 208.980	84 Po (209)	85 At (210)

Oxygen heads Group VIA of the Periodic Table. It is more reactive than nitrogen, its neighbor in Group VA, but less so than fluorine.

Oxygen

Oxygen (O) is the most abundant element on Earth. It makes up about 23 per cent by weight and 21 per cent by volume of the atmosphere, 89 per cent by weight of water, and almost half the weight of the Earth's crust. In air, oxygen is found mainly as a diatomic molecule (O_2); elsewhere it is found combined with many other elements as oxides, in various salts, and as a constituent of living matter.

In the Periodic Table, oxygen lies at the head of Group VIA. By taking up two electrons from other elements it can acquire a full outer shell of electrons and a stable structure. By doing this, it can combine with almost every other element — both metals and nonmetals. Compounds of the other elements and oxygen are known as oxides, and can be formed in a variety of ways. For example, finely divided iron or carbon burn in pure oxygen to form oxides, whereas iron left in the air slowly rusts, also to form iron oxides. Such reactions with oxygen are known as oxidation. Most combustion reactions involve oxidation — in these cases a very fast process accompanied by the evolution of heat and light — as does the decay of plant and animal tissue.

Role of oxygen in biology

Oxygen is a colorless, odorless and tasteless gas which is essential to life. It is used by animals to break down food and provide energy, and is involved in the photosynthesis reactions that occur in the leaves of green plants.

In human beings, the gas is breathed into the lungs where it reacts with the red pigment of blood — the complex molecule called hemoglobin. An unstable compound is formed between hemoglobin and oxygen — known as oxyhemoglobin — and it is this which gives the blood in the arteries its bright red color. Oxyhemoglobin is transferred through the arteries to tissues in the body, where it releases its oxygen to be used in the breakdown of food and waste products in the cells. Complex organic molecules (enzymes) help to speed or catalyze these reactions at body temperatures.

One of the products formed in tissue cells after reactions such as these is carbon dioxide (CO_2). This is carried back to the lungs by the blood, where it is exhaled. When oxyhemoglobin has lost its oxygen it becomes deep purple, and this accounts for the color of venous blood.

In green plants, oxygen is involved in the photosynthesis reactions which occur between carbon dioxide and water. These reactions, stimulated by light, help to form the sugars, starches and cellulose plants need to grow. Because oxygen is slightly soluble in water — it contains about 3 per cent by volume of oxygen — aquatic animals and plants are able to make use of it in much the same way that terrestrial animals and plants use air. Dissolved oxygen also helps to break down sewage and other wastes in natural water to form harmless substances.

Production of oxygen

Oxygen is made in very large quantities by the fractional distillation of liquid air. It is used in the

The Earth's atmosphere contains about 21 per cent (by volume) of oxygen at sea level. The amount of oxygen decreases rapidly with altitude, so that at orbital heights (far right) there is virtually no oxygen at all. Within the stratosphere ultraviolet radiation from the Sun converts some oxygen into its allotrope ozone (mechanism below), which acts as a filter that blocks most ultraviolet radiation.

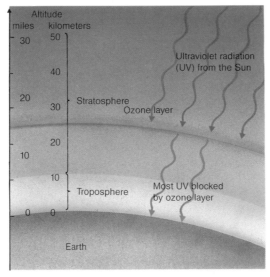

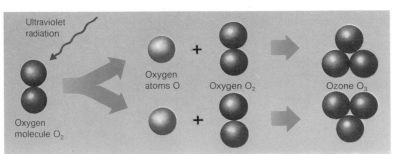

production of steel, the manufacture of chemicals from natural gas, and the formation of important industrial oxygen compounds such as oxirane (C_2H_4O) and sodium peroxide (Na_2O_2).

In the fractional distillation process, air is filtered and dried and carbon dioxide removed. The resulting gas mixture — containing nitrogen, oxygen and the rare gases (mainly argon) — is compressed and cooled until it liquefies. The liquefied air is then distilled in a fractionating column, in which the liquid becomes richer and richer in oxygen and the gas distilled off richer and richer in nitrogen. Liquid oxygen — which is pale blue in color — is removed 99.5 per cent pure and transferred to steel pressure vessels for further use.

Oxygen can also be made by the electrolysis of water. In this process an electric current passed through water yields oxygen at the anode (positive electrode) and hydrogen at the cathode. Both of these gases can be collected.

Allotropes of oxygen

Oxygen exists in two forms (allotropes) with different physical and chemical properties: oxygen itself and ozone (O_3). Ozone is a gas like oxygen but it has a garliclike odor and very different chemical properties. It is produced by passing air or oxygen between two plates that are electrically charged with an alternating potential of several thousand volts. This "silent" electrical discharge produces only a very small amount of the gas, which is diluted with air and oxygen.

Ozone is often also formed in the air during lightning storms in the upper atmosphere, in maximum concentrations at a height of about 15 miles (25 kilometers). This so-called ozone layer is of vital importance in protecting life on Earth from too much exposure to ultraviolet radiation, which it blocks. Fears have been expressed that waste gases from industry and from aerosol sprays are damaging the ozone layer, although recent measurements of ultraviolet radiation in Antarctica suggest that the risk is exaggerated.

Ozone reacts with other elements much more readily than does oxygen, and its enhanced oxidizing properties are finding increasing use. Liquid ozone is dark blue in color and can explode violently, producing ordinary oxygen.

Atomic weights

Since 1961 the atomic weights of the elements have been calculated using the commonest isotope of carbon, ^{12}C, as a reference element. Previously oxygen was used as a reference, mainly

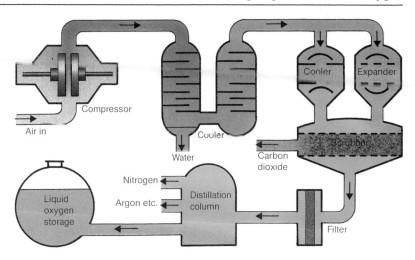

because of the ease with which it combined with many other elements. Oxygen exists in three isotopic species: ^{16}O, which accounts for 99.759 per cent in nature; ^{17}O (0.0374 per cent); and ^{18}O (0.2039 per cent). Chemists used to calculate atomic weights using the naturally occurring mixture of these three isotopes as their reference, whereas physicists used only ^{16}O as theirs, and this led to much confusion. (Now the atomic weight standard is ^{12}C, with an atomic weight of 12.000 atomic mass units.)

The oxygen isotopes ^{18}O and ^{17}O can be obtained by the fractional distillation of water. ^{18}O is used to study the reaction mechanisms of oxygen-containing compounds by detecting its presence using mass spectrometry. Compounds of ^{17}O and ^{18}O can also be prepared and used to investigate metabolic reactions involving oxygen.

The commercial preparation of oxygen involves the liquefaction of air under pressure, which is fractionally distilled to remove nitrogen and argon and other rare gases.

Fishes "breathe" the oxygen in water. Fresh water contains about 3 per cent of dissolved oxygen, which is removed in a fish's gills and passed to the animal's bloodstream.

Fact entries

Oxygen was discovered in the 1770s independently by two chemists, Karl Scheele (1742—1786) in Sweden in 1772 and Joseph Priestley (1733—1804) in England in 1774, who obtained the gas by heating mercuric oxide. But it was the French chemist Antoine Lavoisier (1743—1794) who correctly explained the role of oxygen in combustion. He also coined the name oxygen, from Greek words meaning acid-maker, wrongly believing that all acids contain oxygen.

Oxides can be classified into three main types: acidic, basic and neutral. Acidic oxides react with water to form acids; examples include carbon dioxide, CO_2 (which forms carbonic acid) and sulfur trioxide, SO_3 (which forms sulfuric acid). Basic oxides form bases with water; an example is sodium oxide, Na_2O (which forms sodium hydroxide). Neutral oxides, such as nitrous oxide (N_2O) and water (H_2O), are neither acidic nor basic.

Some metal oxides can be acidic or basic, and are termed amphoteric. For example, aluminum oxide (Al_2O_3) reacts with an acid to form an aluminum salt or with an alkali to form an aluminate salt of the alkali metal.

VA	VIA	VIIA
7 N 14.007	8 O 15.999	9 F 18.998
15 P 30.974	16 S 32.060	17 Cl 35.453
33 As 74.922	34 Se 78.960	35 Br 79.904
51 Sb 121.750	52 Te 127.600	53 I 126.904
83 Bi 208.980	84 Po (209)	85 At (210)

Sulfur to polonium follow oxygen to fill the remainder of Group VIA of the Periodic Table. Tellurium is unusual in having a higher atomic mass than its right-hand neighbor (iodine).

Sulfur to polonium

The elements sulfur (S), selenium (Se), tellurium (Te) and polonium (Po) make up, with oxygen, Group VIA of the Periodic Table. From sulfur to polonium the elements are chemically very similar, and there is a definite gradation of properties down the group. Oxygen has different properties from the rest — primarily because of its higher electronegativity, its ability to attract electrons to itself in a chemical bond to another atom. It is also a gas, whereas the others are solids.

The metallic character of the elements increases from sulfur to polonium. Sulfur is a yellow nonmetal, selenium and tellurium show some metallic properties (tellurium has a silvery luster), and polonium is like lead, being dense and soft with a low melting point. It is also highly radioactive, with a half-life of 138 days, emitting alpha radiation. This makes polonium difficult to handle, and the chemistry of the element and its compounds difficult to study.

Structural forms
Sulfur exists in a number of different forms, or allotropes, in each of the three phases — solid, liquid and gas. Solid crystalline sulfur is made up of puckered rings of eight sulfur atoms and takes one of two crystals shapes: rhombohedral and monoclinic. If sulfur is crystallized from a solution in carbon disulfide (CS_2) below 205°F (96°C), it forms orthorhombic crystals, termed alpha-sulfur or $S\alpha$. If crystallized from a similar solution heated above 237°F (114°C), monoclinic sulfur crystals are formed, termed $S\beta$.

Both allotropes have different arrangements of puckered rings of eight sulfur atoms. An amorphous or plastic form of sulfur, which has no definite crystal structure, can be formed by pouring molten sulfur into cold water. This allotrope consists of long chains of sulfur atoms, and slowly changes to the rhombohedral $S\alpha$ form. Unlike $S\alpha$ and $S\beta$, plastic sulfur is not soluble in carbon disulfide.

Sulfur also shows changes in structure and properties when it is melted. On heating, S_8 sulfur (the sulfur made up of eight-membered rings, either $S\alpha$ or $S\beta$) melts to give a transparent yellow liquid. This turns brown and thickens as the temperature rises above 284°F (140°C), becoming thickest at about 390°F (about 200°C). Above this temperature it starts to thin again. At the boiling point of sulfur — 832.3°F (444.6°C) — it is a dark red, mobile liquid. The change in properties of liquid sulfur results from the breaking open of S_8 rings in $S\alpha$ and $S\beta$ as the temperature rises. In the vapor phase sulfur exists as a mixture of S_8, S_4 and S_2 molecules in equilibrium.

Selenium and polonium also exhibit allotropy. Two unstable forms of selenium, made up of Se_8 rings very similar to S_8 rings, can be obtained by evaporating solutions of selenium in carbon disulfide at about 162°F (72°C). Crystals of the stable gray form of the element can be grown from hot melts. These are unlike any of the sulfur allotropes and are made up from infinitely long spiral chains of selenium atoms. There is a low-temperature allotrope of polonium with a cubic crystal structure, which is stable below 212°F (100°C), and a high-temperature form with a rhombohedral structure. Tellurium exists only in the silvery-white metallic form.

Occurrence
Natural deposits of sulfur occur in many parts of the world associated with salt domes — rock outcrops of sodium chloride and gypsum with veins of sulfur in them — and salt basins — large areas of sedimentary salts containing sulfur. Elemental sulfur, once called brimstone, is also found in the rock deposits left by volcanoes. Much sulfur is extracted from domes and basins using the Frasch process, which is very similar to drilling for oil. Sulfur is also produced from sulfur-bearing natural gas found in some parts of the world and, in Japan, from volcanic rocks.

Selenium and tellurium are not as widespread

The Frasch process is an ingenious method of mining sulfur from deep under the ground. The sulfur occurs in a layer mixed with calcite (calcium carbonate) and is reached by means of a bore hole; three concentric pipes are sunk down the hole. Water superheated to 311°F (155°C) and under pressure is pumped down the outer pipe and compressed air forced down the inner one. The hot water melts the sulfur, which accumulates at the end of the pipes, and a frothy mixture of sulfur, air and water passes up the third pipe to the surface, where it is run off to set in large molds. More than 80 per cent of the world's sulfur is obtained this way in Texas and Louisiana.

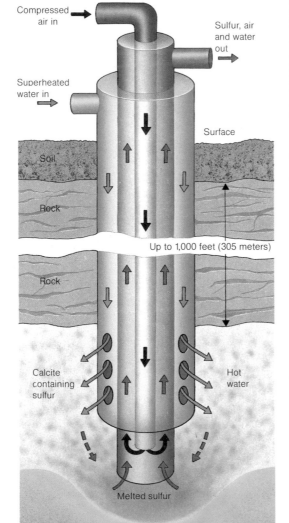

Compressed air in

Sulfur, air and water out

Superheated water in

Surface

Soil

Rock

Up to 1,000 feet (305 meters)

Rock

Calcite containing sulfur

Hot water

Melted sulfur

Anhydrite

as sulfur and are usually found associated with metal sulfide ores — particularly those of silver and gold. Both elements can be extracted from the sludge left after the electrolytic production of copper.

Selenium has unusual physical properties, which can be utilized. Alloys of the element with copper, iron and other metals are used in the manufacture of rectifiers (electrical devices that change alternating current to direct current), photocells and solar cells (devices that produce an electric current when exposed to light), and in photocopying. Tellurium is very similar to selenium in this respect and can be used in the manufacture of light-emitting diodes (LEDs). Some tellurium is also used to make phosphors for television screens.

Polonium occurs as a product of radioactive decay in minerals containing uranium and thorium. It was first isolated from pitchblende — a black mineral often found near silver deposits — by Marie Curie (1867—1934), who also discovered radium. An isotope of polonium, ^{210}Po, can be produced artificially. It is used as a power source in satellites and to eliminate static electricity in some industries.

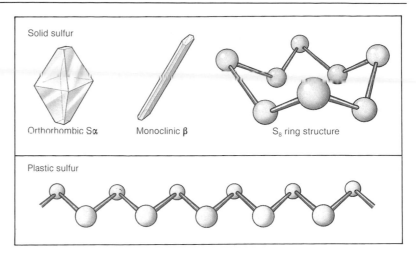

Solid sulfur

Orthorhombic Sα   Monoclinic β   S₈ ring structure

Plastic sulfur

Sulfuric acid

Most sulfur is utilized in the production of sulfuric acid (H_2SO_4), which is made on a vast scale for the production of fertilizers from phosphate rock and ammonia. It is also used in the manufacture of soaps, detergents, plastics and explosives, and in the refining of oil.

Most sulfuric acid is made by the contact process, in which sulfur is first burned in air to produce sulfur dioxide (SO_2). This gas is reacted with more air in the presence of a catalyst to give the higher oxide, sulfur trioxide (SO_3). Sulfur trioxide can combine with water to produce sulfuric acid. But this is not very efficient and better results are obtained if sulfur trioxide is absorbed into sulfuric acid itself, with which it can combine to form an even more concentrated acid. The product of the contact process is usually acid of 98 per cent concentration.

Solid sulfur occurs in two crystalline forms: orthorhombic (Sα, below 205°F, or 96°C) and needle-shaped monoclinic (Sβ, above this temperature to the melting point, 246°F, or 119°C). Both have a molecular structure consisting of a puckered ring of eight sulfur atoms. Molecules of plastic sulfur consist of long zigzag chains.

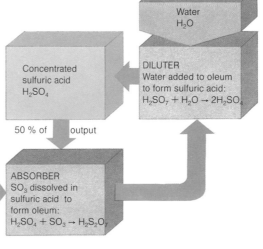

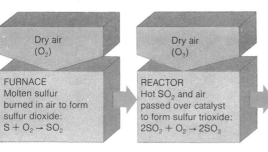

Water
H_2O

Concentrated sulfuric acid H_2SO_4

DILUTER
Water added to oleum to form sulfuric acid: $H_2SO_7 + H_2O \rightarrow 2H_2SO_4$

50 % of   output

Dry air (O_2)

Dry air (O_2)

FURNACE
Molten sulfur burned in air to form sulfur dioxide: $S + O_2 \rightarrow SO_2$

REACTOR
Hot SO_2 and air passed over catalyst to form sulfur trioxide: $2SO_2 + O_2 \rightarrow 2SO_3$

ABSORBER
SO_3 dissolved in sulfuric acid to form oleum: $H_2SO_4 + SO_3 \rightarrow H_2S_2O_7$

Sulfuric acid is probably the most important bulk chemical manufactured in industrial countries. Most is made in stainless steel plants (far left) by the contact process. Sulfur dioxide (SO_2), usually obtained by burning sulfur in dry air, is oxidized to sulfur trioxide (SO_3) using air and a vanadium (V) oxide catalyst. The SO_3 is absorbed in concentrated sulfuric acid to give oleum, which is then diluted with water.

Fact entries

Sulfur, known since ancient times, was first classified as an element by the French chemist Antoine Lavoisier (1743—1794) in 1777. It is named from the Latin *sulphurium,* meaning brimstone. At. no. 16; at. mass 32.060; m.p. 119°C; b.p. 445°C.

Selenium was discovered by the Swedish chemist Jöns Berzelius (1779—1848), who named it after *Selene,* the Greek word for the Moon. At. no. 34; at. mass 78.960; m.p. (gray allotrope) 217°C; b.p. 685°C.

Tellurium was extracted from Transylvanian ores in 1782 by the Austrian geologist Franz Müller von Reichenstein (1740—1825), who named it after the Latin *tellus* meaning earth. At. no. 52; at. mass 127.60; m.p. 450°C; b.p. 990°C.

Polonium was discovered in pitchblende in 1898 by the French scientists Pierre Curie (1859—1906) and Marie Curie (1867—1934), and named after Poland, Marie Curie's homeland. At. no. 84; at. mass (209); m.p. 254°C; b.p. 962°C.

	0
	2 He 4.003
VIA	VIIA

8 O 15.999	9 F 18.998	10 Ne 20.179
16 S 32.060	17 Cl 35.453	18 Ar 39.948
34 Se 78.960	35 Br 79.904	36 Kr 83.800
52 Te 127.600	53 I 126.904	54 Xe 131.300
84 Po (209)	85 At (210)	86 Rn (222)

The halogens, which make up Group VIIA of the Periodic Table, show a graduation of chemical properties from the highly reactive fluorine to the comparatively unreactive iodine.

The halogens

Fluorine (F), chlorine (Cl), bromine (Br), iodine (I) and astatine (At) are five reactive elements commonly called the halogens. They make up Group VIIA of the Periodic Table and have many properties in common. Fluorine and chlorine are gases, bromine is a liquid at ordinary temperatures, and iodine and astatine are solids.

Fluorine, chlorine, bromine and iodine are too reactive to occur alone in nature, and readily form ionic or covalent bonds with other elements. Astatine is radioactive and only small quantities of it have been made.

The halogens readily form negatively charged ions — F^-, Cl^-, Br^- and I^- — by gaining an electron, and are highly electronegative. Fluorine is the most electronegative, and most chemically reactive, of all the elements. It combines with every element except oxygen and the lighter inert gases. Because of the ease with which they form negative ions (anions), the halogens are also the most nonmetallic of the elements. They can even form compounds with themselves, the so-called interhalogens. With the exception of compounds BrCl, ICl, ICl_3 and IBr, they are all halogen fluorides.

Occurrence of the halogens

The elements occur as diatomic molecules. Fluorine and chlorine are greenish-yellow gases under normal conditions, whereas bromine is a dense, mobile, dark red liquid. Iodine is a black solid which has a slight metallic luster. Astatine was discovered in 1940. About 20 isotopes have been found, all of which decay rapidly; the longest-lived decays to half its mass in 8.3 hours.

The elements are found in nature, combined with other elements, as minerals and salts (the name halogen comes from the Greek words meaning salt-formers). Fluorine and chlorine are the most abundant. Fluorine occurs in fluorite (CaF_2, calcium fluoride) and in the aluminum ore cryolite (Na_3AlF_6). The most common source of chlorine is the sodium chloride (NaCl, common salt) found in seawater and inland salt lakes. Bromine is also found in seawater and in brines as the magnesium and sodium bromides associated with the corresponding chlorides. Iodine occurs in brines in a similar state to bromine; it also concentrates in some sea plants. Astatine was first discovered as the product of a nuclear reaction of an isotope of bismuth.

Extraction and uses of fluorine

Early attempts to produce fluorine from aqueous solutions containing salts of the element failed because of fluorine's high electronegativity. Now it is made by electrolysis of molten potassium hydrogen fluoride (KHF_2) in steel or copper vessels. Fluorine is usually kept in metal containers because, although there is an initial reaction with the metal, a protective film of metal fluoride often forms.

Because it is so reactive, fluorine itself is of little use. Not only do most elements react with it but some organic compounds catch fire in the gas. Fluorine compounds are, however, widely used. Hydrofluoric acid (HF) is used as a catalyst in the petroleum industry. The acid and its salts are used to etch glass, particularly in the manufacture of "frosted" electric light bulbs. Polytetrafluoroethylene (PTFE) is an inert plastic which can withstand large differences in temperature. Freon-12 (CF_2Cl_2, a chlorofluorocarbon) is a refrigerant

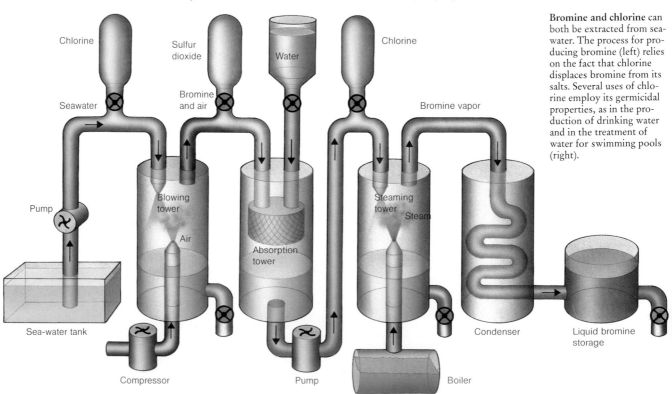

Bromine and chlorine can both be extracted from seawater. The process for producing bromine (left) relies on the fact that chlorine displaces bromine from its salts. Several uses of chlorine employ its germicidal properties, as in the production of drinking water and in the treatment of water for swimming pools (right).

Chlorine • Sulfur dioxide • Water • Chlorine

Seawater • Bromine and air • Bromine vapor

Pump • Blowing tower • Air • Absorption tower • Steaming tower • Steam • Condenser • Liquid bromine storage

Sea-water tank • Compressor • Pump • Boiler

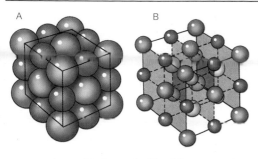

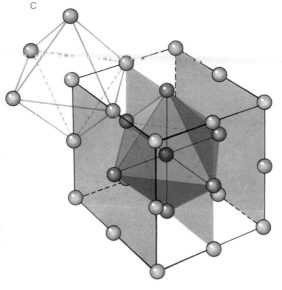

Sodium chloride crystals form a lattice of sodium ions (Na^+) and chloride ions (Cl^-) held together by electrostatic attraction — ionic bonding. The structure is a face-centered cube (A) with ions at each corner and at the centers of each face (B). The ion at the center (C) — sodium or chloride — is closest to eight ions of the other type arranged octahedrally around it.

commonly used in household refrigerators. Some freons have also been used as propellants for aerosols.

Until comparatively recently the rare gases helium, neon, argon, krypton, xenon and radon were considered to be chemically inert; this is why they were once called the inert gases. In 1962 though, it was found that compounds could be formed between fluorine and xenon. Three xenon fluorides have been prepared — XeF_2, XeF_4 and XeF_6 — as well as complex ions containing both elements. Xenon chloride ($XeCl_4$) is now known to exist as well as krypton fluoride (KrF_2), which is a white, volatile solid.

Extraction and uses of chlorine

Most chlorine is prepared on a large scale by the electrolysis of brine — a process which gives both the gas and sodium hydroxide — and it is also made as by-product in the electrolytic extraction of metals from their molten chlorides. Large quantities of the element are used for bleaching textiles and paper, and in the manufacture of bleaching powder and household bleach. Chlorine is used to kill bacteria in the domestic water supply and in swimming pools.

Hydrogen chloride (HCl) is produced from salt for the manufacture of hydrochloric acid (a

solution of hydrogen chloride in water). This strong acid dissolves many metals to give their chlorides. Electrolysis of concentrated hydrochloric acid can be used to produce chlorine.

Extraction and uses of bromine and iodine

Bromine is used in the production of fumigants for seeds and soil, and in the petroleum industry. It can be prepared quite easily from brines by acidifying with sulfuric acid (H_2SO_4) and adding chlorine. Bromine is formed in the solution and can be swept out in a current of air. Some bromine compounds are used to manufacture dyes and photographic emulsions, and organic compounds containing the element are useful fire-extinguishing and flame-proofing chemicals.

Iodine is also obtained from brines by methods similar to those used for bromine extraction. The element is a solid at room temperature, but when heated it sublimes (vaporizes without melting) to give a violet-colored vapor. Iodine dissolves only slightly in water, but is soluble in many organic solvents in which it forms brown or purple solutions. It is thought that dimerization occurs to some extent in these solutions, with I_4 and I_2 molecules existing in equilibrium.

Compounds of iodine have been widely used as disinfectants and germicides, and in the preparation of drugs and dyestuffs. Iodine is present in the human thyroid gland, which is located at the base of the neck. Insufficient iodine in the diet to maintain the supply to the thyroid causes it to swell and leads to the condition known as goiter.

Bluejohn, seen here carved into an ornate vase, is a unique banded form of the mineral fluorite (calcium fluoride, CaF_2) that is found only in Derbyshire, England.

Fact entries

Fluorine was first isolated in 1886 by the French chemist Henri Moissan (1852—1907). It occurs mainly in the mineral fluorite, after which it is named. At. no. 9; at. mass 18.998; m.p. —219.6°C; b.p. —188°C.

Chlorine was discovered by the Swedish chemist Karl Scheele (1742—1786) in 1774. It was named later after its greenish-yellow color (*chloros* is Greek for green). At. no. 17; at. mass 35.453; m.p. —101°C; b.p. — 34°C.

Bromine, named from *bromos,* the Greek for stench, was discovered in 1826 by the French chemist Antoine Balard (1802—1876), who extracted it from the ashes of seaweed. At. no. 35; at. mass 79.904; m.p. —7.2°C; b.p. 58.8°C.

Iodine was discovered in seaweed ash by the French chemist Bernard Courtois (1777—1838) in 1811. It is named after the violet color of its vapor (Greek *iodes* means violet). At. no. 53; at. mass 126.904; m.p. 113.5°C; b.p. 184.4°C.

Astatine is an artificial element, first prepared at the University of California by Emilio Segrè and others in 1940. Its 20 or so isotopes are radioactive; it was named after the Greek *astatos* meaning unstable. At. no. 85; at. mass (210).

	0
VIIA	2 **He** 4.003
9 **F** 18.998	10 **Ne** 20.179
17 **Cl** 35.453	18 **Ar** 39.948
35 **Br** 79.904	36 **Kr** 83.800
53 **I** 126.904	54 **Xe** 131.300
85 **At** (210)	86 **Rn** (222)

The rare gases occupy Group O of the Periodic Table. They react only with fluorine and chlorine, halogens from Group VIIA.

A neon sign consists of a glass tube (A) containing neon gas at low pressure. Positive ions, striking an electrode, generate secondary electrons. A neutral neon atom (B) has ten electrons. A secondary electron excites an outer electron (C), making it jump to a higher energy level. When the excited electron jumps back, its excess energy is emitted as red light (D).

The rare gases

The rare gases form a group of six elements that are remarkable for their unreactivity. For this reason known also as the noble gases, they are helium (He), neon (Ne), argon (Ar), krypton (Kr), xenon (Xe) and radon (Rn). Their unreactivity — a reluctance to enter into chemical combination with other elements to form compounds — is a consequence of their electronic structure. Atoms of elements that have full outer electron shells have no strong attraction for electrons from other atoms, nor do they readily give up electrons to other atoms. All the rare gases have such full outer shells.

Compounds and uses

The rare gases lost their claim to total inertness in the early 1960s, when several compounds of krypton, xenon and radon were prepared by treating them with the highly electronegative halogen elements fluorine and chlorine.

This does not mean, however, that the elements themselves are not of value. The argonons, as they are sometimes called, have become important industrial gases, often because of their unreactivity. Electric light bulbs, for example, are usually filled with a mixture of 90 per cent argon and 10 per cent nitrogen. The inert "atmosphere" in the bulb prevents metals atoms evaporating from the incandescent filament. As a result, modern lamps have a much longer working life than earlier ones, which contained a vacuum that allowed such evaporation, resulting in rapid thinning and weakening of the filament.

Argonons are used to provide inert atmospheres in metallurgy, chemical processing and nuclear reactors. In each application, the rare gas blankets a material that would react violently if exposed to air (and would be adversely affected by nitrogen, the most commonly used "blanket").

The most familiar use of the argonons is probably in a neon sign. When an electric current is passed through a tube containing neon at a low pressure, the ionized gas emits a red light. By using various gas mixtures, it is possible to produce the wide range of colors of so-called "neon" signs. Xenon is the gas in the tube of a photographer's electronic flash gun, and in the high-power electric lamps used in some lighthouses.

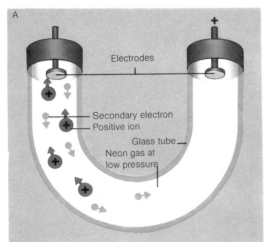

A Electrodes / Secondary electron / Positive ion / Glass tube / Neon gas at low pressure

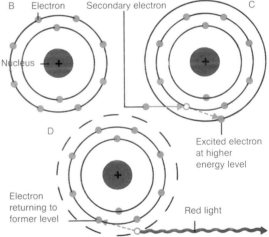

B Electron / Secondary electron / Nucleus / C / D / Excited electron at higher energy level / Electron returning to former level / Red light

The AD500, a modern semirigid airship seen here flying over the Eiffel Tower, Paris, is filled with helium gas, which is second only to hydrogen in lightness. Instead of venting (and therefore losing valuable gas) to reduce buoyancy in order to descend, the AD500 uses engine-driven ducted fans to drive the ship downward.

Helium is less dense than air and is used to inflate balloons and airships as an alternative to highly inflammable hydrogen. Safer diving systems have also been developed using helium. Breathing ordinary compressed air causes nitrogen to dissolve in the blood, which can give divers "the bends" if they surface too rapidly. Artificial air made by mixing helium and oxygen overcomes this problem; the only side effect is the squeaky "Donald Duck" speech of someone breathing the mixture.

At extremely low temperatures ($-454°F$, or $-270°C$) helium liquefies. If cooled even further, its properties suddenly change; for example, it flows up the inside wall of its container and down the outside (superfluidity). More importantly, if certain metals and alloys are cooled to liquid helium temperature, they too show new properties, notably a loss of electrical resistance. This phenomenon (superconductivity) is exploited in the magnets for giant particle accelerators.

Radon is radioactive, but has a very short half-life and breaks down in a few days, with the emission of alpha particles (helium nuclei). It has been used in medicine for treating tumors. But it can harm as well as help. Radon is liberated into the air in mines where the rocks contain radioactive materials. There has recently been concern about the health hazard that this can present to miners.

Discovery and extraction

Many of the rare gases occur in minute amounts in air — hence the name "rare." The first hint that they existed came in the late 1700s, when Henry Cavendish obtained a small residue of unknown gas after taking a sample of air and removing from it all the oxygen and nitrogen. More than a century later, Lord Rayleigh and William Ramsay discovered argon (using a spectroscope) while trying to account for the different densities of nitrogen extracted from air and nitrogen prepared from its chemical compounds. They went on to discover that air also contains three other argonons, and today neon, argon, krypton and xenon are obtained by the distillation of liquid air. In 1868 helium had been detected in the spectrum of the Sun — it was later found in uranium ores (1895) and as a component of volcano gases (1898) and natural gas (1903) — and radon was discovered in 1900 as a decay product of the radioactive element radium.

Colorful neon signs are used for advertising throughout the world.

Welding of aluminum is carried out under a "blanket" of argon gas. The inert argon does not react with the hot metal, nor does it allow oxygen in the air to oxidize it.

Fact entries

Helium was discovered in uranium ore in 1895 by the British chemist William Ramsay (1852—1916). It had been named after the Greek *helios* meaning Sun, because it was first detected in the Sun's spectrum. At. no. 2; at. mass 4.003; b.p. $-268.9°C$.

Neon was also discovered by Ramsay, with Morris Travers (1872—1961), in 1898 in impure argon from air. Its name derives from the Greek *neos*, meaning new. At. no. 10; at. mass 20.179; b.p. $-246°C$.

Argon was isolated from air in 1894 by Ramsay and Lord Rayleigh (1842—1919). It was named from the Greek *argos*, meaning inactive. At. no. 18; at. mass 39.948; b.p. $-185.9°C$.

Krypton, named after the Greek *kryptos*, meaning hidden, was discovered by Ramsay and Travers in 1898. At. no. 36; at. mass 83.80; b.p. $-153.4°C$.

Xenon, from the Greek word for strange, was also discovered by Ramsay and Travers in 1898. At. no. 54; at. mass 131.30; b.p. $-108°C$.

Radon was discovered by the German chemist Friedrich Dorn (1848—1916) in 1900, and later isolated by Ramsay. It is formed by the radioactive decay of radium, after which it is named, and is itself radioactive. At. no. 86; at. mass (222); b.p. $-62°C$.

IIA		
4 Be 9.012		
12 Mg 24.305	IIIB	IVB
20 Ca 40.080	21 Sc 44.956	22 Ti 47.900
38 Sr 87.620	39 Y 88.906	40 Zr 91.220
56 Ba 137.340	57 La 138.906	72 Hf 178.490
88 Ra (226)	89 Ac (227)	104 (261)

The scandium group makes up Group IIIB of the Periodic Table. It includes the lanthanides (elements 57 to 71) and the actinides (elements 89 to 103).

Many of the phosphors, which produce the colors on television screens and on VDU displays for computer graphics, contain lanthanide compounds.

The scandium group and the lanthanides

Scandium (Sc), yttrium (Y), lanthanum (La) and actinium (Ac) are four elements which have certain physical and chemical properties in common. They form Group IIIB of the Periodic Table. However lanthanum and actinium have their own series of related elements, known as the lanthanides and the actinides, which form separate groups within the Periodic classification. The lanthanides are included in this article, and the actinides are the subject of the next one.

In some respects, the Group IIIB elements resemble those of Group IIIA: boron, aluminum, gallium, indium and thallium. The resemblance is particularly strong with scandium, whose chemistry is much like that of aluminum. All the Group III elements form triply charged positive ions, the size of which increases from boron to thallium and on to scandium through to lanthanum. The sizes of the ions of yttrium and lanthanum are more similar to those of the other lanthanide elements, and therefore their chemistry is more closely associated with the lanthanides.

Occurrence
Scandium is quite common; there is almost as much of the element in the Earth's crust as there is arsenic and almost twice as much as there is boron. But it is not separated easily from other elements and there are few rich mineral sources. Scandium occurs mixed with other lanthanides and is separated using special cation exchange processes — which involve the exchange of positive ions (cations) of the element in solution with ion-exchange resins.

Yttrium is a heavy element resembling lanthanum and the lanthanides. Its ions are similar in size to those of terbium (Tb) and dysprosium (Dy), and it is found associated with them.

Rare earths
The lanthanides (or lanthanons) used to be known as the rare earth elements because they were found only rarely, and then only as oxides or "earths." But by modern standards they are not rare and some are produced on a large scale. Even the scarcest — thulium (Tm) — is as common as bismuth and occurs in more abundance than arsenic, mercury, cadmium or selenium.

Apart from promethium (Pm), the lanthanides always occur together — most commonly in the mineral monazite, a dense, dark sand which contains mixtures of lanthanide phosphates. Lanthanum, cerium, praseodymium and neodymium make up to 90 per cent of all mineral deposits of lanthanides, with yttrium and the heavier elements of the group making up the rest. Promethium does not occur naturally; several artificial isotopes have been made, but only ^{145}Pm and ^{147}Pm, which is produced in very small quantities as a product of nuclear fission, have half-lives of more than two years.

Extraction and separation
The lanthanides are separated from the other elements by reaction with nitric acid (HNO_3), and precipitation as oxalates or fluorides. They are then separated from each other using large-scale ion-exchange techniques, after the recovery of

Monazite, a phosphate mineral which is the principal source of cerium and other lanthanides (and of thorium), occurs in beach sands along the southwestern coast of Pakistan.

cerium and europium compounds by further precipitation. As the solution containing a mixture of lanthanide ions passes through a column of ion-exchange resin, the ions are preferentially absorbed by the resin in a specific order: lutetium ions (the smallest) are absorbed first and lanthanum ions (the largest) are absorbed last. Thus there is usually a separation into zones along the resin in the column, with each zone containing ions of a particular element. The ions are removed by "washing" the resin in the column with various solutions, and they are obtained as ionic species complexes.

The metals from lanthanum to gadolinium can be made by reacting lanthanum trichloride ($LaCl_3$) — or similar compounds for the other elements — with calcium metal at about 1,800°F (about 1,000°C). The others, and yttrium, can be made by a similar method using the trifluorides.

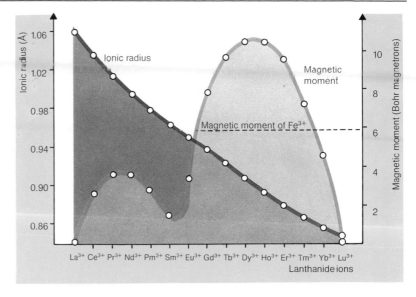

Lanthanide contraction

The lanthanide elements are silvery-white metals, which are chemically very reactive. They tarnish rapidly in air when not protected and burn easily to give oxides. All react with water to give hydrogen. In contrast, yttrium does not burn in air until it is heated to about 1,800°F (about 1,000°C), because of the formation of an oxide coating on the surface when it is exposed to the air. Lanthanides easily give up three electrons from their outermost electron shells to form stable positive ions (cations) with three positive charges — they are all electropositive elements. The ions formed get smaller from La^{3+} to Lu^{3+}. This shrinkage, known as the lanthanide contraction, accounts for many of the chemical properties of the elements and is a consequence of the complicated outer electronic structures of the atoms (usually there is an increase in ionic size with increasing atomic number). The other physical properties of the lanthanides change fairly smoothly along the period, including the volume of the atoms and the heat needed to vaporize them. There are, however, sudden changes at the elements europium (Eu) and ytterbium (Yb). These two metals are untypical lanthanides: they can form stable divalent (not trivalent) cations, and this affects their properties.

Color

The lanthanides are similar to the transition metals in that many of their compounds are colored and exhibit a characteristic magnetic behavior. The lanthanides are different, however, because of the larger number of electrons surrounding their nuclei. Most lanthanide compounds have pale colors — lilac, primrose, pale green and pink, for example — as opposed to the brighter compounds of the other transition elements such as iron and cobalt. Some chemicals containing lanthanide ions also fluoresce and luminesce — they emit visible, usually colored, radiation when excited by light or charged particles such as cathode rays. These effects are widely used in phosphors for color television screens.

The lanthanide contraction (above) refers to the steady decrease in size of the ions along the series (rather than the expected increase). All except the first and last are paramagnetic, with some having higher magnetic moments even than iron (Fe^{3+}).

Element	Symbol	Discovery date	Discoverer
Cerium	Ce	1803	Jöns Berzelius (1779—1848) and Martin Klaproth (1743—1817)
Praseodymium	Pr	1885	Carl von Welsbach (1858—1929)
Neodymium	Nd	1885	Carl von Welsbach (1858—1929)
Promethium	Pm	1945	J. A. Marinsky *et al.*
Samarium	Sm	1879	Lecoq de Boisbaudran (1838—1912)
Europium	Eu	1896	Eugène Demarçay (1852—1904)
Gadolinium	Gd	1880	J. de Marignac (1817—1894)
Terbium	Tb	1843	Carl Mosander (1797—1858)
Dysprosium	Dy	1886	Lecoq de Boisbaudran (1838—1912)
Holmium	Ho	1878 1879	J. L. Soret (1827—1890) and Per Cleve (1840—1905)
Erbium	Er	1843	Carl Mosander (1797—1858)
Thulium	Tm	1879	Per Cleve (1840—1905)
Ytterbium	Yb	1878 1907	J. de Marignac (1817—1894) George Urbain (1872—1938) and
Lutetium	Lu	1907 1908	George Urbain (1872—1938) and Carl von Welsbach (1858—1929)

Fact entries

Scandium, one of the "missing" elements predicted by Dmitri Mendeleyev, was discovered in 1879 by the Swedish chemist Lars Nilson (1840—1899). He named it after *Scandia,* the Latin name for Scandinavia. At. no. 21; at. mass 44.956; m.p. 1,538°C; b.p. 2,832°C.

Yttrium is one of several elements named after the town of Ytterby in Sweden. It was isolated by the French chemist George Urbain (1872—1938) in 1907 and independently a year later by the Austrian Carl von Welsbach (1858—1929). At. no. 39; at. mass 88.906; m.p. 824°C; b.p. 1,193°C.

Lanthanum was discovered in 1839 as an impurity in cerium by the Swedish chemist Carl Mosander (1797—1858) and named from the Greek *lanthanein,* which means to be hidden. At. no. 57; at. mass 138.906; m.p. 920°C; b.p. 3,454°C. Brief details of the discovery of the various lanthanides or rare earth elements are given in the above table. Actinium is described in the article which follows.

The actinides and beyond

The elements from actinium (Ac, atomic number 89) to lawrencium (Lr, atomic number 103) are known as the actinides. They form a series which is an offshoot of Group IIIB of the Periodic Table, and have many physical and chemical characteristics in common. All of the actinides are radioactive and beyond uranium (U, atomic number 92) they do not usually occur naturally. Only minute amounts of elements 93 and 94, neptunium (Np) and plutonium (Pu), have been found in nature and all the elements beyond these have been made only synthetically.

Uranium is extracted in large quantities for use as a fuel in nuclear reactors; neptunium and plutonium are formed as a result of the radioactive decay of this uranium fuel and are recovered from it. Beyond fermium (Fm, atomic number 100), all the elements are only very short-lived, rapidly undergoing radioactive decay to other elements. The most stable isotopes of mendelevium, nobelium and lawrencium (Md^{258}, No^{255} and Lr^{256}) have half-lives of only 53 days, 185 seconds and 45 seconds, respectively.

Beyond lawrencium, the last of the actinides, three further elements may have been identified. Each is very short-lived and highly radioactive, and only a few atoms of them have been observed. The names rutherfordium (Rf) and hahnium (Ha) have been proposed for elements 104 and 105, and these are thought to have similar properties to hafnium and tantalum in the third transition series of the Periodic Table. In theory, the number of elements and the periodic classification could extend indefinitely, and so there is a continuous

search for new elements. But as the elements get larger they are likely to get more unstable, undergoing radioactive decay very rapidly. It has been calculated that the element with atomic number 110 would have a half-life of about 10^{-10} seconds (one ten-thousand millionth of a second).

Radioactive emission
The chemical and physical properties of the elements can be explained in terms of their electronic structure, or the energy levels in which the atoms' electrons exist around the nucleus. The periodic classification brings together elements with similar electronic structures, and therefore similar properties, into groups and periods. Trends in physical and chemical properties can be followed throughout the classification, and these have been used in the past to predict the existence of elements before they were actually discovered. For some elements though, the nucleus of the atom begins to play an important part in determining the properties of the element. This is particularly the case with the elements from actinium onward, which are all radioactive. The nucleus of the atoms of these elements is continually undergoing change — the neutrons seem to be less effective in counteracting the repulsive electrostatic forces between the protons, and to reach a stable condition they emit radioactive particles.

By emitting an alpha or beta particle the nucleus of the atom of one element changes and becomes an atom of another element. An alpha particle consists of a helium nucleus ($_2He^4$), and so alpha emission results in a loss of 4 units of

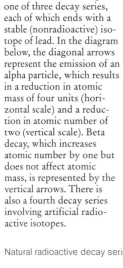

Naturally occurring radioactive elements belong to one of three decay series, each of which ends with a stable (nonradioactive) isotope of lead. In the diagram below, the diagonal arrows represent the emission of an alpha particle, which results in a reduction in atomic mass of four units (horizontal scale) and a reduction in atomic number of two (vertical scale). Beta decay, which increases atomic number by one but does not affect atomic mass, is represented by the vertical arrows. There is also a fourth decay series involving artificial radioactive isotopes.

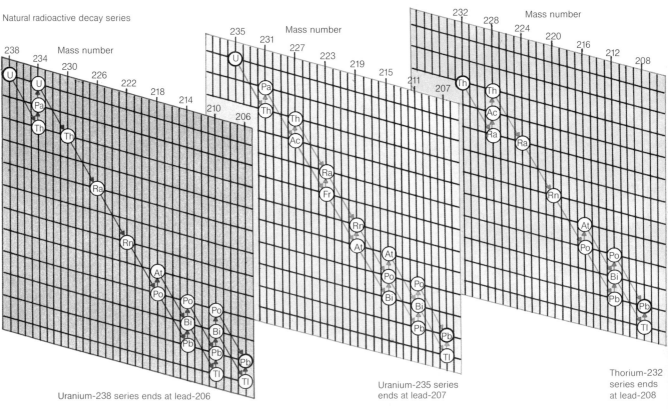

Natural radioactive decay series

Uranium-238 series ends at lead-206

Uranium-235 series ends at lead-207

Thorium-232 series ends at lead-208

mass (2 protons and 2 neutrons) and 2 units of positive charge, resulting in an element two places farther down the periodic classification.

Loss of a beta particle (an electron) results in no change in mass number, but forms an element one place higher in the Periodic Table. Thus alpha decay of the radium isotope Ra^{226}, for example, results in Rn^{222} (radon), and beta decay of Ra^{228} forms Ac^{228} (actinium).

If the immediate product of radioactive decay is itself radioactive, it also decays to form another element. There may be a whole series of such transitions. The uranium isotope U^{238}, for example, decays spontaneously by the emission of both alpha and beta particles through a series of intermediates to a stable isotope of lead, Pb^{206}. This radioactive decay involves the formation of 16 isotopes other than U^{238} and Pb^{206}, including the uranium isotope U^{234} and the thorium isotopes Th^{234} and Th^{230}.

All the elements from actinium onward are continually undergoing such changes at different rates, producing different isotopes with different half-lives. For the elements with higher atomic numbers these half-lives are short, and the radioactive decay is rapid. For some of the elements with lower atomic numbers, however, such as thorium and protactinium, the half-lives are much longer (in the range of many thousands or millions of years). Th^{232} for example occurs naturally and has a half-life of 1.39×10^{10} years (13,900 million years).

The actinide metals

Although the actinides and the transition elements beyond lawrencium are highly radioactive, their chemistry has been studied (depending on the availability of the element), and some trends emerge. All are metals, showing physical and chemical metallic characteristics, and the size of the tripositive ions from actinium (Ac^{3+}) to americium (Am^{3+}) has been shown to decrease — as happens passing along the lanthanide series from lanthanum to lutetium. Actinium is a silvery-white substance which glows in the dark because of its radioactivity. Otherwise its chemistry, both as a solid and in its soluble compounds, is very similar to that of lanthanum. Thorium is again a white metal but it tarnishes rapidly in the air. It has been machined and forged but is very much like the lanthanide metals in being highly electropositive — it is able to lose electrons very easily, and catches fire in air if in finely divided form.

Protactinium is an unreactive metal which tarnishes in air. Uranium also tarnishes, first to give a yellow and then to a black film which does not protect the bulk of the metal. Neptunium is a silvery metal, like uranium, as is plutonium. Plutonium is also a very toxic substance, used as a fuel for nuclear reactors and in the manufacture of nuclear weapons. One gram of Pu^{238} produces 0.56 watts of thermal power, which results mainly from alpha particle emission. This emission was used in the Apollo space project to power experiments left on the surface of the Moon.

Scientific experiments left on the Moon by Apollo astronauts were powered by a thermoelectric generator which made use of the heat produced by the radioactive decay of an isotope of plutonium (Pu^{238}).

A geologist (left) uses a geiger counter to detect radioactivity in rocks while prospecting for uranium minerals.

Fact entries

Radioactivity The German physicist Röntgen first observed radioactivity in 1895 when he noticed that a cathode-ray tube emitted rays that could penetrate substances which were opaque to visible radiation. A year later, the French physicist Henri Becquerel found that uranium compounds affected a photographic plate even if it was covered with paper or a thin sheet of metal. He found that the intensity of the rays, which he called radioactivity, was proportional to the amount of uranium present in the compound. Marie and Pierre Curie investigated the effects of the radioactivity of uranium, and discovered the elements polonium and radium, from which Friedrich Dorn discovered radon.

Half-life of a radioactive element is the time taken for half the atoms of the element present at any instant to disintegrate. Its amount and radioactivity halve in this time. Half-lives can be of the order of a few thousand million years or of a few hundred-thousandths of a second. The rate of disintegration of a radioactive element depends only on the structure of its nuclei, and is not affected by temperature, pressure or chemical combination.

Metals and alloys

Metals are characterized by their density, strength, thermal and electrical conductivity, and the ease with which they can be formed into different shapes. In many applications, however, not all of these properties are needed — or wanted — at the same time. Sometimes hard metals are required, sometimes soft ones. Often they need to show great strength, or resistance to chemical attack. And sometimes no single metal has exactly the required combination of properties.

To provide materials to suit such a wide range of applications, metallurgists have developed alloys. An alloy consists of two or more elements, at least one of which is a metal, and is formed by deliberately mixing the so-called base metal with an alloying element or elements to give the required degree of strength, formability, and so on. Alloys have a myriad of applications, from high-strength and high-resistance steels to light magnesium-based alloys used in the construction of aircraft. In practice very few metals are used in the pure state.

The importance of alloys
The significance of alloys in our development is

illustrated by the great leap forward that civilization made — in southeastern Asia in about 4000BC and in Europe about 2,000 years later — when it was found that mixing two soft and easily formed metals, copper and tin, resulted in a material that could be worked easily but was much harder than either of the two starting materials. This alloy, bronze, gave its name to the Bronze Age, and was used for the manufacture of weapons and tools which were widely traded. Bronze is still used today for making machine parts and statues. Today's bronzes contain about 90 per cent copper and 10 per cent tin, with traces of other elements added to give a range of properties. Adding small amounts of zinc and lead gives an easily-machined bronze known as gun metal.

Properties of metals and alloys
The properties of alloys depend on both the atomic structure and microscopic structure of metals themselves. Most elements used in alloys are metals, in which the atoms are arranged in a regular pattern or crystal lattice. Each atom has a stable core of electrons surrounding its nucleus, which contains all the electrons except the outer-

Plan view of one layer of metal atoms

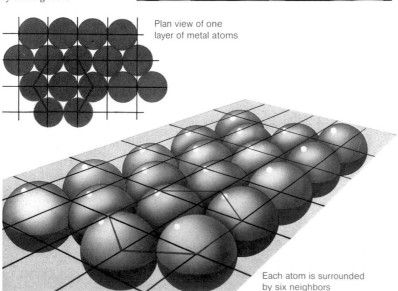

Each atom is surrounded by six neighbors

The hexagonal close-packed lattice

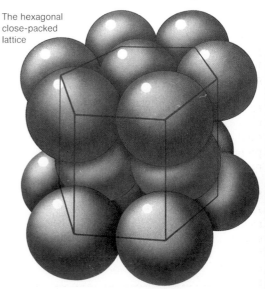

most valence electrons. These outer electrons can be shared between neighboring atoms and form a "sea" of electrons, which is distributed evenly throughout the metal and attracts the positively charged atomic cores. The attractive forces are not in any particular direction, and give rise to a very strong form of bonding known as metallic bonding. This accounts for the special properties of metals. Electrons can be made to flow easily through the electron sea if a potential difference (voltage) is applied to the metal; thus metals are good conductors.

Because of the closely packed and regular arrangement of the atoms and their core electrons — each atom can be surrounded by up to 12 others with 6 more atoms not too far away — metals are often very dense. And because the metal atoms are attracted equally in all directions, metals can easily be formed into different shapes. The close-packed and regular arrangement of metal atoms allows them to form crystals of a variety of types, depending on the atomic arrangement. If molten metal is cooled slowly, comparatively large crystals form; if cooled rapidly, a glassy or amorphous type of metal can be formed, with no definite crystal structure. In practice, metals are usually a mixture of the two types.

Interstitial and substitutional alloys
At the atomic level, alloys are of two distinct types known as interstitial alloys and substitutional alloys. Interstitial alloys are formed when atoms of a small nonmetallic alloying element (hydrogen, boron, carbon or nitrogen, for example) are mixed with the base metal. They become trapped in the spaces, or interstices, in the metal's atomic lattice. Alloys of this type are known for their hardness.

In substitutional alloys, metallic elements with atoms of a similar size to those of the base metal are mixed with it. Atoms of the alloying element and the base metal can interchange and replace each other in the base-metal lattice. The most important alloys of this type are the so-called alloy steels. In these, iron atoms are interchanged with those of other transition metal elements such as chromium, manganese or nickel. Brass is also an interstitial type alloy — a mixture of copper and 18 to 30 per cent zinc — as are the alloys used to make coins, such as cupronickel (copper and nickel) and cuprogold (copper and gold).

In practice, many alloys are a mixture of both types — interstitial and substitutional. This is particularly the case with ordinary steels, which often contain very small proportions of silicon, phosphorus and sulfur as impurities and the major alloying elements.

Solid solutions
Alloying can be compared with dissolving a solute in a solvent — dissolving sugar in water for example. An alloy is therefore really a solid solution, and the base metal is said to possess solid solubility. As with liquid solutions, an excess of solute added to the solvent causes precipitation. It

is sometimes possible to cool alloys very quickly so that expected precipitates do not form. Such alloys are said to be in a metastable state, and have many useful properties.

Also, precipitates formed in alloys may be compounds of the base metal and the alloying element. Tungsten carbide is an example, formed between the metal tungsten and carbon in a tungsten/carbon alloy. It has distinct metallic properties, as do similar compounds of metals and carbon or nitrogen (nitrides) often found in alloys. Deliberate deposition of these so-called intermetallic compounds provides metallurgists with the possibility of changing the structure and properties of alloys to confer upon them particular required properties.

The versatility of modern metals is demonstrated by the table setting (above) in which everything — including the flowers and cloth — is made from steel.

A modern trader in Oman, whose forebears probably sold earthenware pots, displays a wide variety of aluminum and stainless steel vessels.

Commercial electroplating makes extensive use of complex compounds, from the small batch plating of silverware (above) to the automated line for brass plating hundreds of small handles (above right).

Complex inorganic compounds

Many inorganic compounds are not simple two-element substances such as carbon dioxide (CO_2) and sodium chloride (NaCl). They consist of a number of elements bonded together in a complicated structure. Compounds in which a central atom is surrounded by other atoms, ions or small molecules are known as complexes or coordination compounds.

Complexes can be electrically neutral molecules or electrically charged ions — complex ions — and usually have from two to nine atoms, ions or groups surrounding a central atom or ion. Some of the many hundreds of thousands of complexes and complex ions have important applications, ranging from helping to carry oxygen from

the lungs and through the blood of animals, to developing images on photographic films.

Ligands and the coordinate bond

The atoms, ions or small molecules surrounding the central atom or ion of a complex are known as ligands, and the number of them is called the coordination number. Such numbers can run from 2 to 9 and in some cases beyond, but the most important complexes in chemistry are those with coordination numbers of 2, 4 or 6. For example, most metal ions dissolved in water form complex ions with a coordination number of six: $[Cu(H_2O)_6]^{2+}$, $[Al(H_2O)_6]^{3+}$ and $[Fe(H_2O)_6]^{3+}$.

The formation of complexes is one of the most outstanding properties of metal ions, and the changes in their properties has given a valuable insight into the metals themselves and the nature of the chemical bond. Complexes of the metals of the first transition series (scandium to zinc) with ammonia for example — ammine complexes such as $[Cu(NH_3)_6]^{2+}$ — have been extensively studied. Some of these metals also form a wide range of complexes and complex ions containing carbonyl (CO) groups as ligands, which have special bonding characteristics.

Any atom, ion or molecule with two extra electrons, known as a lone pair, which it can donate to a central atom can form the ligand of a complex. Ligands which bond through one lone pair are known as monodentates, but many molecules have more than one atom with a lone pair and can function as polydentate ligands. Ammonia is a monodentate ligand because it possesses only one lone pair, whereas the compound 1,2-diamino-ethane (ethylenediamine, $NH_2CH_2CH_2NH_2$) is a bidentate ligand — because it can form two coordinate bonds (one at each end of the molecule) to make a sort of bridge in a complex. In both cases the lone pair of electrons possessed by the ligands originates from the nitrogen atom(s).

Most complexes have characteristic shapes, related to the coordination number of the "central" atom (usually a metal). The most common coordination numbers — 2, 4 and 6 — give rise to molecules with the shapes shown.

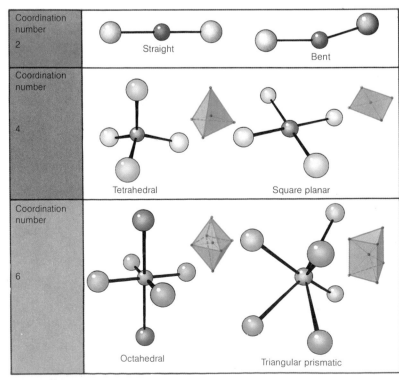

Coordination number 2	Straight	Bent
Coordination number 4	Tetrahedral	Square planar
Coordination number 6	Octahedral	Triangular prismatic

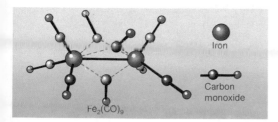

Fe₂(CO)₉

Iron

Carbon
monoxide

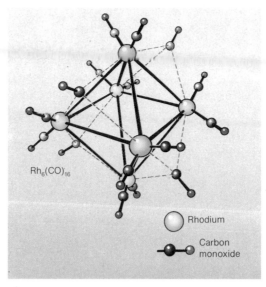

Rh₆(CO)₁₆

Rhodium

Carbon
monoxide

Cluster compounds can have complicated molecular structures, from the relatively simple iron complex $Fe_2(CO)_9$ (far left) to the compound of rhodium $Rh_6(CO)_{16}$, which has 16 carbon monoxide ligands, 12 bonded directly to the six rhodium atoms (disposed at the corners of an octahedron) and four triply-bonded to three rhodium atoms.

The carbonyl group (CO) also acts as a monodentate ligand, like ammonia, but in a special way. The carbon of the carbonyl group is a poor donor atom and usually CO would not be expected to act as a ligand. But as a result of its bonding with oxygen, it is able to draw electrons from the central (metal) atom and donate a pair of electrons to the metal simultaneously. This gives rise to a so-called synergic bond, which is very stable.

To describe the bonding in complexes, a number of theories have been put forward, such as the valence-bond theory (developed extensively by the American chemist Linus Pauling), the crystal field theory, the ligand field theory and the molecular orbital theory. The latter three are particularly useful in describing the shapes, color and magnetic properties of the complexes.

Shapes of complexes

The shapes of complexes and complex ions has been the subject of much study. In complexes with the most common coordination numbers (2, 4 and 6), eight different spatial arrangements of atoms, ions and molecules are found. Complexes with a coordination number of 2 can be straight or bent, whereas complexes with coordination number 4 usually take on a tetrahedral arrangement, although a flat square of atoms or groups is sometimes found (a square planar arrangement) if all the ligands are the same. Complex compounds with coordination number 6 take on an octahedral shape, which may be distorted. A triangular prismatic shape also occurs, but less frequently.

When a complex has two or more different ligands attached to the central atom it may have more than one structural formula. The two different structural arrangements of the same compound are called stereoisomers. When the stereoisomers are mirror images of each other, they are known as optical isomers (because their solutions rotate the plane of polarized light).

Polynuclear complexes

Some metals form compounds in which two or more atoms of the same metallic element are bonded together. The most common examples are complicated structures, much like traditional metal complexes but containing two or more central metal atoms and surrounding ligands which are often carbonyl (CO) or nitrosyl (NO) groups. An example of such a polynuclear complex is $Fe_2(CO)_9$.

With more than two metal atoms in the complex, cluster compounds are formed — for example the rhodium compound $Rh_6(CO)_{16}$. Its structure is quite complicated. Rhodium metal atoms are arranged at the corners of a regular octahedron with carbonyl ligands attached to them. Some of the bonds between adjacent rhodium atoms are further strengthened by bridges of carbonyl groups.

There are other types of complex inorganic compounds that resemble the polymers found in organic chemistry. Solid sulfur trioxide (SO_3), for example, forms as icelike crystals at room temperature which consist of S_3O_9 molecules formed into rings. It is unstable, however, and slowly changes into a stable form with fibrous, needle-like crystals which resemble those of asbestos. They are made up from infinitely long chains of sulfur tetroxide (SO_4) tetrahedra linked by sharing oxygen atoms at two of their corners. The silicate minerals also consist of long chains and rings of atoms. These are formed, in a way similar to solid sulfur trioxide, from arrangements of the orthosilicate ion (SiO_4^{4-}).

Human blood — supplies of which are vital for transfusions in modern medicine — contains hemoglobin, a complex compound whose molecules consist of proteins coordinated to a central iron atom. Similar complexes are essential to many other biological systems.

Organic chemistry

There are more substances in which carbon is the predominant element than those of all the other elements together. The basic reason is the unique ability of carbon to form chemical bonds with both itself and a very large number of other elements. Most important is the ability of carbon atoms to link into long chains or complex cyclic structures. This ability is of profound importance to us. Without it, life on Earth would probably never have arisen. All the fundamental groups of substances found in living organisms — such as proteins and carbohydrates — are based on carbon-atom chemistry.

At one time it was believed that compounds found in living organisms could not be made in the laboratory. Scientists thought that they could be obtained only from matter which was living or had been alive. This idea was shown to be wrong more than 150 years ago. Nevertheless, the branch of chemistry concerned with carbon compounds is still called organic chemistry.

Organic chemistry now embraces not only substances produced by living organisms, but also an immense range of synthetic chemicals, including many of industrial importance. Nearly all plastics and synthetic fibers in everyday use are organic chemicals, as are such diverse materials as dry-cleaning fluids, artificial sweeteners, pesticides and many pharmaceuticals.

Many of today's industrially important organic chemicals are produced from fossil fuels, mainly oil and natural gas. These are the remains of once-living matter so, in that respect, much of industrial organic chemistry retains its truly organic origins. But such is the sophistication of modern chemistry that if supplies of these raw materials become exhausted, then the same products could still be produced from alternative carbon sources, such as carbon monoxide gas.

Carbon's combining power

In its outer shell, a carbon atom has four electrons; the shell is exactly half-filled. A single carbon atom can thus form four two-electron bonds. Not only is it able to form bonds with other carbon atoms, it can also bond readily with hydrogen atoms and with atoms of the elements to its right in the Periodic Table: nitrogen and oxygen. A very large number of organic substances are made up of the elements carbon, hydrogen, oxygen and/or nitrogen. Other elements often found in organic compounds include sulfur, phosphorus and the halogens (particularly chlorine).

In the laboratory in the past few decades, chemists have managed to combine carbon atoms with atoms of most other elements. A whole new branch of chemistry, called organometallic chemistry, has opened up to study the behavior of substances involving carbon-metal bonds. Many of these have found important uses in industry as catalysts. They have also proved valuable in extending scientific understanding of the nature of chemical bonding.

Another significant aspect of carbon-based chemistry is the ability of carbon atoms to link together by more than a single bond. Two carbon atoms can be joined by two or even three pairs of electrons, forming double or triple carbon-carbon bonds. Multiple bonds make molecules containing them reactive in a variety of ways. In simple compounds such as ethene (ethylene), the increased reactivity is put to great use in making a wide variety of industrially important chemicals. In more complex organic substances, such double or triple bonds can contribute to the color of a substance. For example, part of the visual process in human beings depends on a double bond in the compound rhodopsin, which occurs in the retina of the eye.

The following articles demonstrate the build-up of organic chemicals from the most simple — small molecules containing only carbon and hydrogen — through to much more complex ones. In many organic compounds, particular combinations of atoms are found, such as

The various types of plastics (examples of which are shown below) are extremely versatile synthetic organic chemicals which, since their development in the early part of the century — originating from the discovery of Bakelite in 1909 — have transformed modern life to an almost inconceivable extent; it is probably true to say that there is not a single moment in a person's life when he is not using plastics or relying on them indirectly.

—O—H (the hydroxyl group). Much of organic chemistry can be described in terms of the effects that these different functional groups have on the molecule, both physical effects and chemical effects.

Improving on nature

Ultimately, we reach the very complex organic chemicals, such as those used as drugs or pesticides. In these compounds, several different functional groups may occur together. Many have provided extraordinarily complex challenges to the synthetic chemist. Yet many are produced naturally by plants and animals with such efficiency that it is still more economic to extract them from a natural source rather than to make them in the laboratory.

On the other hand, by studying the structures of natural substances, it has been possible in many instances to improve on them and obtain products even better suited to our needs. The study of penicillin, for example, originally a natural substance obtained from a mold, has led to the development of a whole family of compounds. Although the core of the molecule remains the same, one part of it can be changed in various ways to produce different penicillins that are suitable for different kinds of therapeutic treatment. Similarly, the naturally-occurring insecticidal substance in pyrethrum flowers has been modified to give a range of synthetic pyrethroids, with a greater variety of pesticidal uses.

Oil and natural gas, obtained by drilling (above), are the chief starting materials for a wide range of important organic chemicals, such as plastics, pesticides, dyes, synthetic textiles and some pharmaceuticals.

The foxglove plant *Digitalis purpurea* (left) was for many years the principal source of the drug digitoxin, used to treat heart conditions such as cardiac failure. Today this drug has generally been replaced by digoxin, obtained from another species of foxglove, *Digitalis lanata.* Many other drugs and useful organic chemicals are still obtained from plants or animals, usually because the substances cannot be synthesized economically but sometimes because chemists have not yet succeeded in synthesizing them at all.

that share a pair of electrons, one provided by each atom, and is termed a covalent bond. The removal of one hydrogen atom from methane forms a methyl group (denoted by $-CH_3$). Because of the ability of carbon atoms to form chains, two methyl groups can join to form a compound that consists of two carbon atoms and six hydrogen atoms. This substance, called ethane (formula C_2H_6), is the second member of the alkane series.

Just as ethane can be derived from methane, so can one of the hydrogen atoms from the ethane molecule be replaced by a methyl group to form a chain of three carbon atoms with eight hydrogen atoms attached — the compound called propane (C_3H_8). Its middle carbon atom is linked to two other carbon atoms — and thus has room for only two hydrogen atoms.

Isomerism
By repeating the process of replacing a hydrogen atom with a methyl group, the alkane series can be further extended to butane (C_4H_{10}), pentane (C_5H_{12}), hexane (C_6H_{14}) and even larger molecules.

Once an alkane molecule has four or more carbon atoms, it need no longer have the shape of a simple unbranched chain. Instead of adding another methyl group to the end of protane's three-carbon chain, for example, the methyl group can be attached to the middle carbon atom, to form a branched chain. The central atom is linked to three other carbon atoms and only one hydrogen atom; each of the other three carbons has three hydrogens attached.

This branched-chain molecule is called isobutane or 2-methylpropane (indicating that the additional methyl group is attached to the second carbon atom). The formula of isobutane and ordinary straight-chain butane (called n-butane, where n stands for normal) is C_4H_{10}. Both forms have the same numbers of carbon and hydrogen atoms as each other, but their structures are different.

This phenomenon, in which molecules have the same chemical composition but a different structural arrangement of atoms, is called structural isomerism, and the different forms are called isomers. The physical properties of structural isomers can vary widely; for example, n-butane boils at a temperature of 31°F (−0.5°C) whereas isobutane boils at 10.9°F (−11.7°C).

Production and uses
The first few members of the alkanes — from methane to butane — have important domestic and industrial uses. They are obtained mainly from fossil fuels, occurring either as underground fields of natural gas or as gas associated with deposits of oil. In some areas, such as under the North Sea, natural gas is almost pure methane; elsewhere — in the United States, for example — it may contain a significant proportion of other hydrocarbons, such as ethane, propane and butane. The longer-chained alkanes — gasoline

Natural gas, a major source of alkanes, is often burned off as a waste product at oil drilling installations (above). Increasingly, however, the gas associated with oil deposits is not wasted: it may be piped ashore from offshore fields; transported as a liquid in refrigerated ships; or converted to methanol for transport over longer distances.

Saturated aliphatic hydrocarbons

The saturated aliphatic hydrocarbons form a series of organic molecules that contain only hydrogen and carbon (hence the term hydrocarbon). Each carbon atom is linked to four other atoms — the maximum number possible — and is therefore called saturated. And in every member of the series the molecules are arranged in the shape of straight or branched chains (aliphatic), rather than in closed rings. Formerly known as paraffins, the members of this series are now called alkanes.

The alkanes
The first member of the alkane series and the simplest organic compound is methane, the chief component of natural gas. A molecule of methane consists of a single carbon atom linked to four atoms of hydrogen; its chemical formula is CH_4. Each of its chemical bonds involves two atoms

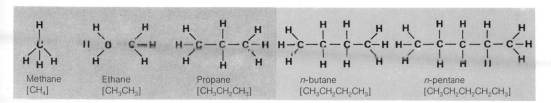

Methane [CH₄] Ethane [CH₃CH₃] Propane [CH₃CH₂CH₃] n-butane [CH₃CH₂CH₂CH₃] n-pentane [CH₃CH₂CH₂CH₂CH₃]

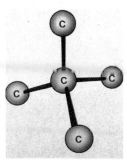

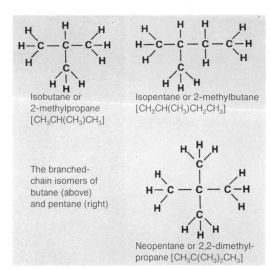

Isobutane or 2-methylpropane [CH₃CH(CH₃)CH₃]

Isopentane or 2-methylbutane [CH₃CH(CH₃)CH₂CH₃]

The branched-chain isomers of butane (above) and pentane (right)

Neopentane or 2,2-dimethylpropane [CH₃C(CH₃)₂CH₃]

The first five straight-chain members of the alkanes are illustrated above, with their formulas in square brackets. Each member differs from its immediate neighbors by one carbon atom and two hydrogen atoms, a relationship that is reflected in the general formula for alkanes: C_nH_{2n+2} (where n is the number of carbon atoms). Only one structure — a straight chain — is possible for the first three alkanes; but with four or more carbon atoms, different, branched-chain structures (which have the same general formula as the straight-chain alkanes) are also possible. The alternative forms (called isomers) of butane and pentane are illustrated on the right. Butane and pentane have only one and two branched-chain isomers respectively, but the number increases dramatically as the number of carbon atoms increases: a 40-carbon alkane has more than 62 million isomers.

Each carbon atom can link to four other atoms (including other carbon atoms). When this happens, the atoms tend to arrange themselves so that they are an equal distance apart, with the result that the bonds are directed towards the corners of a regular tetrahedron — as illustrated in the diagram above, which shows the structure of diamond, and also as exemplified by the structure of methane (far top left).

(petrol), kerosene (paraffin), oils, greases, waxes and bitumen — are obtained by distillation of petroleum.

Methane is used mainly as a fuel, although it is also an important chemical feedstock for the manufacture of other chemicals. Methane can, for example, be converted into methanol (CH_3OH) and other, longer-chained alcohols as well as into long-chain alkanes such as gasoline. Methane can also be used to make trichloromethane (chloroform, $CHCl_3$), and tetrachloromethane (carbon tetrachloride, CCl_4).

The principal use of ethane is as a chemical feedstock, chiefly to make ethene (ethylene, C_2H_4), which can itself be converted into other useful substances — the plastic polyethylene, for example.

Propane and butane can easily be converted into liquids and stored in pressurized containers, to be used as a portable fuel supply. They are also used to make other chemicals, such as ethene and propene (propylene, C_3H_6).

Syngas

It is estimated that, at the present rate of comsumption, known reserves of natural gas will last only until about the year 2030. For this reason, there has been considerable research into developing ways of treating coal (reserves of which are estimated will last several hundred years) to make a synthetic gas — called syngas — similar to natural gas.

When any carbon-containing material is burned in a limited amount of air, carbon monoxide is formed; this gas can be reacted with hydrogen to produce methane. Scientists have developed various ways of making syngas from coal and other organic material but, as yet, none is commercially feasible.

Biogas

Methane is also generated by some living organisms as a by-product of their metabolism. In developing countries, such as India and China, farmers build "digesters" to make methane from waste organic matter, such as animal dung. In the absence of air, bacteria feed on the waste and give off methane. The local people use the gas for heating and lighting. Such biogas systems are also being investigated in developed countries, principally as a productive means of disposing of large amounts of waste from the intensive farming of animals.

Liquefied alkanes are a convenient form of portable fuel supply; butane, for example, is used in cigarette lighters. Camping stoves and lamps, and even automobiles and trucks can also be fueled by liquid alkanes — called liquid petroleum gas (LPG) in automotive applications.

Alicyclic hydrocarbons — the first four of which are illustrated in the diagram (top right) — resemble both straight- and branched-chain alkanes in consisting of only carbon and hydrogen atoms. But because their carbons are linked in a ring structure, alicyclic hydrocarbons have a different general formula from that of their chain-structure analogues: C_nH_{2n} (where n is the number of carbon atoms). The three-dimensional shapes of alicyclic hydrocarbons are often depicted as flat rings, but in fact the rings are puckered in all except cyclopropane. This puckering is exemplified well by cyclohexane, which can exist in several puckered conformations, two of which are illustrated (lower right). Any one cyclohexane molecule can be flexed into different conformations, from the "chair" form through various intermediates to the "boat" form and back again. But not all of these forms are equally stable. The chair form is the stablest; at any given moment about 99.9 per cent of cyclohexane molecules are in this form.

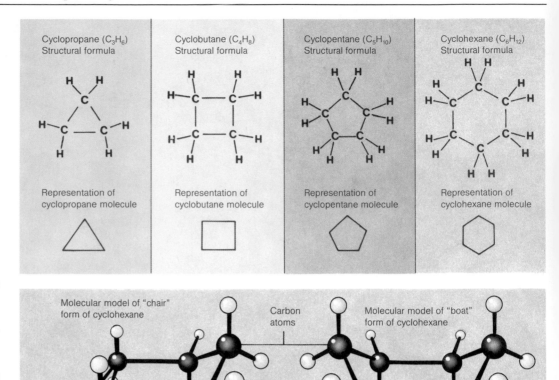

Cyclopropane (C_3H_6)
Structural formula

Representation of cyclopropane molecule

Cyclobutane (C_4H_8)
Structural formula

Representation of cyclobutane molecule

Cyclopentane (C_5H_{10})
Structural formula

Representation of cyclopentane molecule

Cyclohexane (C_6H_{12})
Structural formula

Representation of cyclohexane molecule

Molecular model of "chair" form of cyclohexane

Carbon atoms

Molecular model of "boat" form of cyclohexane

Hydrogen atoms

Cyclohexane molecules twist into different forms, such as the "chair" form (above left, with its representation left) and the "boat" form (above right, representation right)

Reactions of alkanes

The name paraffin, as the alkanes were originally called, derives from the Latin *parum affinis,* meaning little affinity. In chemical terms, the hydrocarbons are relatively unreactive. All, however, undergo oxidation — if they did not, they would not have come to prominence in applications which depend on their combustion.

They also undergo other types of reactions, such as halogenation (combining with one of the halogen elements — fluorine, chlorine, bromine or iodine) and nitration (in which a nitro group $-NO_2$ replaces a hydrogen atom), but the conditions required to sustain these reactions are often severe.

Also under extreme conditions, hydrocarbons will isomerize; that is, they break down partly and then recombine into different hydrocarbon isomers. Such isomerization is at the basis of thermal cracking processes used to increase the octane number of liquid hydrocarbons for gasoline. The octane number is based on the combustion properties of the eight-carbon, straight-chain molecule *n*-octane, which is assigned a value of 100. Most gasoline used today has an octane number between 95 and 98.

As the molecular weight of hydrocarbons increases, they change from gases to liquids to solids at room temperature. Petroleum jelly is a soft, solid hydrocarbon mixture obtained from crude oil; it is used as a barrier ointment to protect wounds and sores. Higher molecular weight hydrocarbons form hard waxes, which can be used as protective coatings or as slow-burning materials — in candles, for example.

Cyclic hydrocarbons

In the simplest hydrocarbons, such as most saturated aliphatics, carbon atoms are joined together in a chain. But because of the unique ability of carbon atoms to combine with one another, it is possible for the two ends of a chain to be linked, to form a "ring" or alicyclic compound.

Clearly, the simplest hydrocarbon which can link together in this way is one containing three carbon atoms: cyclopropane (C_3H_6). It is highly reactive because of the great strain placed on its chemical bonds by the ring structure — the angle between the bonds which link the carbon atoms in cyclopropane is about 60°, only slightly larger than half the 109° angle favored by a carbon atom in its usual tetrahedrally-bonded state.

As the number of atoms in the ring increases, the strain lessens. Consequently, cyclobutane (C_4H_8) is more stable than cyclopropane, but less stable than cyclopentane (C_5H_{10}). The six-carbon ring compound, cyclohexane (C_6H_{12}), is very stable and is frequently used as a chemical reaction solvent. Many of the cyclic hydrocarbons have similar properties to the straight-chain hydrocarbons. Thus, cyclohexane is a colorless liquid at room temperature, is inflammable and will not mix with water — properties shared by its straight-chain analogue n-hexane (C_6H_{14}).

Many of the alicyclic hydrocarbons are found as constituents in crude oil. Some cyclohexane is obtained commercially from fractional distillation of gasoline, but most is obtained by hydrogenation of the aromatic hydrocarbon benzene. Like most other organic compounds, cyclohexane has to be handled with care. A major disaster occurred in Flixborough, England, in 1974 at a chemical plant where cyclohexane was being oxidized as one step in the nylon-making process.

Long-chain hydrocarbons

Very large hydrocarbon molecules can be made, with chains of hundreds or even thousands of carbon atoms. The well-known plastic polyethylene is a synthetic example of such a hydrocarbon. Chains this long do not often occur in nature, although natural rubber is made up of very large molecules formed by the linking together of thousands of molecules of the five-carbon, branched-chain hydrocarbon isoprene ($CH_2=C(CH_3)CH=CH_2$).

The isoprene unit is also the basis for a very widespread group of natural products, called terpenes. These include a large number of natural flavors and fragrances, such as limonene, found in citrus fruits, and alpha-pinene, found in turpentine. These compounds are often complex, sometimes involving ring structures, sometimes having a double bond between one or more pairs of carbon atoms. Often, there are one or more oxygen atoms somewhere in the molecule as well.

Because of their isoprenoid origin, terpenes nearly always have a multiple of five carbon atoms in their structure. Those with 10 carbon atoms are known as monoterpenes, those with 15 as sesquiterpenes. Terpenes with 20 and 30 carbon atoms respectively are known as di- and triterpenes. Tetraterpenes (with 40 carbon atoms) include some important biological materials, which give color to vegetables such as carrots and tomatoes. One such, beta-carotene, can be broken down in the liver to the diterpene derivative vitamin A, which plays an important role in night vision.

The color of tomatoes is due to the presence of a red tetraterpene compound which, like other tetraterpenes, consists of eight linked isoprene units. This structure is similar to that of vitamin A, a substance which the body can produce by breaking down certain types of tetraterpenes.

The nylon-making plant at Flixborough, England, was devastated in 1974 when cyclohexane exploded as it was being oxidized (one stage in the synthesis of nylon 6). Like other alkanes, cyclohexane can be oxidized completely to carbon dioxide and water, a reaction that releases a large amount of energy if it is not controlled.

Ethene (ethylene, C_2H_4)	Propene (propylene, C_3H_6)	Propadiene (C_3H_4)
H—C=C—H (with H's)	H—C—C=C (with H's)	H—C=C=C (with H's)
1-butene (C_4H_8)	cis-2-butene (C_4H_8)	1,3-butadiene (C_4H_6)
structure	structure	structure
trans-2-butene (C_4H_8)	2-methylpropene (C_4H_8)	1,2-butadiene (C_4H_6)
structure	structure	structure

Alkenes are unsaturated aliphatic hydrocarbons with one carbon-carbon double bond; the first few members of the series are illustrated above left. Ethene (in the pink box) and propene (pale blue box) each have only one possible structure, but butene has four (in the pale green panel). Alkadienes have two carbon-carbon double bonds; the first few members are illustrated above right. Propadiene (in the grey box) has only one possible form, whereas butadiene (dark green box) has two.

Unsaturated aliphatic hydrocarbons

Unsaturated aliphatic hydrocarbons are the building blocks of the organic chemical industry. The term "unsaturated" indicates that the molecules contain at least one carbon-carbon multiple bond. This may be a double or a triple bond — compounds with double bonds are also commonly known as alkenes or olefins. By far the most important alkene is ethene (ethylene, C_2H_4), followed by propene (propylene, C_3H_6) and

the diolefin (or alkadiene) 1,3-butadiene ($CH_2CHCHCH_2$, usually called simply butadiene). Ethyne (acetylene, C_2H_2) is the most important alkyne, as compounds with a triple carbon-carbon bond are called.

Compounds with two or more multiple bonds are termed polyunsaturated (a characteristic of many natural fats and oils). Where there are two double bonds in an alkene, they may be between successive carbons (C=C=C) or between alternate pairs (C=C—C=C), when they are said to be conjugated, as in 1,3-butadiene ($H_2C=CH-CH=CH_2$). The color of many organic compounds depends on the arrangement of double bonds in the molecule.

Reactions of unsaturated compounds

In general, multiple bonds are more reactive than are primary carbon-carbon bonds. Compounds containing multiple bonds may undergo polymerization, cyclization and addition reactions. Polymerization is the most important of these — hydrocarbon polymers include plastics, resins, and both natural and synthetic rubbers. In a cyclization reaction, a straight chain curls round to form a ring — several important solvents are made in this way. Addition reactions break one of the bonds in a multiple bond and add two new groups onto the molecule, one at each end of the broken bond. It is this reactivity of the multiple bond, as well as the multitude of synthetic possibilities, that makes these compounds so important in the chemical industry.

One consequence of multiple bonding is that the two linked carbon atoms are not able to rotate freely, as they would do if joined by a single bond. In the case of an alkene (olefin), this leads to cis-trans isomerism. If an alkene has two different atoms or groups attached to each carbon atom, it can exist in two physically different forms. If the

Car tires contain a large proportion — more than 25 per cent (by weight) in some cases — of carbon black, which greatly increases their elasticity, strength and resistance to wear. The carbon black is produced by burning ethyne (acetylene, C_2H_2) in a limited supply of air.

Ethyne (acetylene, C_2H_2)	Propyne (C_3H_4)	1-butyne (C_4H_6)	2-butyne (C_4H_6)	Butadiyne (C_4H_2)
$H-C\equiv C-H$				$H-C\equiv C-C\equiv C-H$

same groups are on the same side of the molecule, the compound is the *cis* isomer. If the same groups are on opposite sides of the molecule, then the compound is the *trans* isomer.

Uses of alkenes

Simple alkenes are produced by cracking naphtha (a straw-colored liquid obtained during the refining of crude oil) or from some types of natural gas. Cracking is the process used in the petrochemical industry to break big molecules down into smaller ones. Naphtha, a mixture of saturated hydrocarbons comprising mainly molecules with 4 — 12 carbon atoms, undergoes thermal cracking at 1,000—1,200°F (about 540—650°C) in the absence of air. This breaks it down into products such as ethene, propene and butadiene. Although natural gas is primarily methane, some sources (notably in the USA) contain ethane, which can be converted into almost pure ethene.

As well as being the single most important organic chemical feedstock, ethene is also an important natural product, involved in the ripening of fruits, for example. It is a colorless, highly inflammable gas with a sweet odor and taste. About half of the ethene (ethylene) produced industrially is used to make polyethylene. Many other plastics, synthetic rubbers and resins can also be derived from it. Ethene is used to synthesize ethylene glycols (used as antifreeze and in cosmetics, paints and lacquers) and ethanoic acid (acetic acid, a major constituent of vinegar). Ethyne can also be made from it.

Propene (propylene) is a colorless, inflammable gas which can be used to produce isopropyl alcohol (2-propanol, C_3H_8O), from which is made acetone (2-propanone, C_3H_6O), an important organic solvent. Cumene (1-methylethylbenzene, C_9H_{12}), produced from propene and benzene, is used to make phenol and acetone. Propylene glycol is a solvent for fats, oils, resins, perfumes, colors and dyes, soft drink syrups and flavor extracts. Butadiene is important in the manufacture of synthetic rubbers and ABS (acrylonitrile-butadiene-styrene) resins. These are extremely hard, durable and resistant to fire.

Higher alkenes can be produced by thermal cracking of waxes or by building up a long chain from ethene. Either method produces a mixture of olefins with 4—20 carbon atoms. Once these have been separated, they are used to make detergents, polymer additives and lubricants.

Acetylene and the alkynes

Ethyne (acetylene) has the formula C_2H_2 and the structure $H-C\equiv C-H$. A colorless, poisonous, highly inflammable gas, it is used for welding because it burns with an intense, hot flame. Carbon black, used as a pigment in typewriter ribbons and carbon paper as well as in tires and plastics, consists of fine particles of carbon produced by burning ethyne in a limited supply of air. Many of the early synthetic rubbers, such as Neoprene and Buna, were developed from ethyne.

Ethyne is produced mainly by the cracking of oil, but it can also be obtained from the reaction between calcium carbide and water:

$$CaC_2 + H_2O \rightarrow Ca(OH)_2 + C_2H_2$$

(calcium carbide is made by roasting coke with calcium oxide, or quicklime).

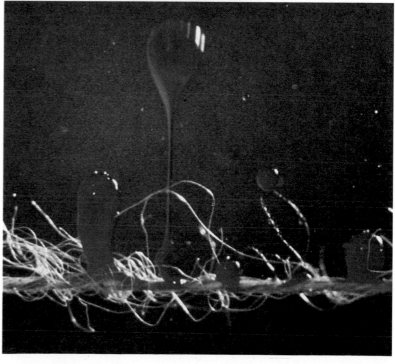

Alkynes (above left) are unsaturated aliphatic hydrocarbons that contain a carbon-carbon triple bond. Because this type of bond is relatively unstable, alkynes are highly reactive. Alkadiynes, such as butadiyne (above right), contain two such bonds and are therefore even more reactive.

Many types of cosmetics, including theatrical make-up, contain ethylene glycols, which can be synthesized from the alkene ethene (ethylene, C_2H_4).

Red oil droplets are floating off a fiber strand (below) as a result of the action of a detergent in the liquid. Many detergents are synthesized using long-chain alkenes (with between 4 and 20 carbon atoms) as starting materials.

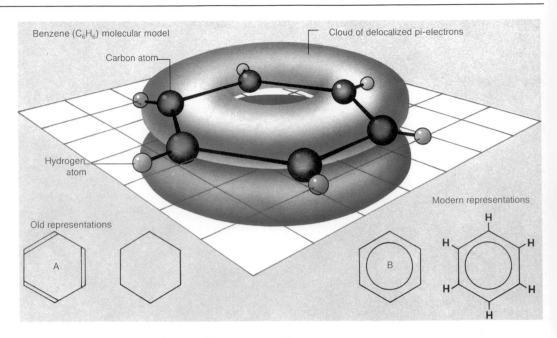

Benzene (C_6H_6) molecular model

Carbon atom

Cloud of delocalized pi-electrons

Hydrogen atom

Old representations

A

Modern representations

B

Benzene is the simplest aromatic compound, and it is usually the presence of at least one benzene ring that characterizes other compounds as aromatic. The structure of benzene (right) is unusual because, instead of having alternate long single and shorter double carbon-carbon bonds in the six-membered ring, all the bonds are of equal length. The "spare" pi-electrons (that would otherwise be needed for the double bonds) are spread evenly round the ring as two doughnut-shaped clouds above and below it. Representations of the structure include the old Kekulé formula (A) and the modern Robinson formula (B).

Aromatic hydrocarbons

All aromatic hydrocarbons are composed solely of carbon atoms and hydrogen atoms in various arrangements, but what sets most of them apart from other hydrocarbons is the presence in their molecular structures of at least one benzene ring.

A benzene ring consists of six carbon atoms linked together in a planar hexagonal structure. In benzene itself, the building block from which aromatic chemistry sprang, each carbon is also bonded to a single hydrogen atom. It is the ability of chemists to replace these hydrogen atoms with other groupings that has given rise to the vast number of aromatic hydrocarbons we know today. Many also occur naturally in the coal and oil deposits which lie beneath the Earth's crust, and others are widespread in nature itself.

Some of the most commonly used aromatic hydrocarbons include toluene and the xylenes. Every year chemists isolate millions of tons of these compounds from petroleum (crude oil) and, to a lesser extent, from coal. They are important industrially, especially as solvents and as starting materials for the manufacture of plastics. They are also used by laboratory chemists for conversion into other, more valuable compounds.

Benzene
Known since 1825, benzene is a colorless liquid at room temperature and, like many of its derivatives, is one of the chief constituents of coal tar. It can also be isolated from petrochemical feedstocks such as naphtha by catalytic reforming, or by extraction from pyrolysis gasoline. Where possible, chemists use distillation methods to separate the various aromatic hydrocarbons, because each compound boils off from the coal tar mixture at a different temperature. But some have boiling points so close together that satisfactory separation by distillation is impossible. Chemists then use solvents to dissolve the required products selectively. Aromatic hydrocarbons are isolated from coal tar and from petrochemical feedstocks using combinations of these techniques.

In the second half of the nineteenth century, the potential utility of benzene derivatives became evident. Scientists noted the presence of the benzene nucleus in many naturally-occurring drugs and dyes, and searched for ways to make them synthetically. But chemists investigating benzene were puzzled by its structure until as late as the 1920s, and many models were put forward

In oil refineries, such as the one shown below, a combination of fractional distillation and selective dissolving techniques is used to isolate aromatic hydrocarbons from petrochemical feedstocks. Fossil fuels, principally oil and coal, are the chief sources of aromatic hydrocarbons.

to explain its molecular makeup. Nowadays, its structure has been rationalized using the concepts of delocalized electrons and resonance forms.

Structures of most organic molecules are drawn in terms of single, double or triple bonds between carbon atoms, each atom possessing the ability to form four covalent bonds. But the properties and reactions of benzene do not conform with those of any single structure drawn in this conventional way and so instead, the six carbon atoms in benzene are considered to be held together by two-electron bonds of equal length and the six electrons which would otherwise be required for the three double bonds are evenly distributed around the hexagonal ring. These six delocalized electrons, known as pi-electrons, give rise to a doughnut-shaped electron cloud distribution above and below the plane of the ring atoms. As a result, all of the bonds between carbon atoms in benzene are the same length. They are shorter than a carbon-carbon single bond in a simple aliphatic compound but longer than an ordinary carbon-carbon double bond.

The greater mobility of the pi-electrons accounts for the unique properties of benzene and its derivatives. For example, the reaction between bromine and a simple aliphatic alkene, such as ethene (ethylene), is an addition reaction — the molecule of bromine adds to the doubly-bonded carbon atoms. The reaction between bromine and benzene, however, is a substitution, with one of the hydrogen atoms being replaced by a bromine atom.

Since its discovery, benzene has fulfilled many roles industrially. It is an excellent solvent and, by suitable use of reagents, can be converted into a vast range of derivatives. Styrene, a compound formed by adding benzene to ethene, is the sub-

unit from which polystyrene is made, and many other plastics, rubbers and resins are built from benzene derivatives. It is also the precursor for many substances essential to the synthetic chemist. Among the better-known benzene derivatives are nitrobenzene, aniline, phenol, benzaldehyde, benzoic acid and chlorobenzene.

Despite its undoubted usefulness, benzene must be treated with caution because it is a carcinogen. As such, it has been banned from the classroom in most Western countries, and is used with great care in the laboratory and in industry.

Toluene

A liquid at room temperature, toluene is identical to benzene except that one of the hydrogen atoms has been replaced with a methyl group (i.e. it is methylbenzene). Like benzene, from which it is easily synthesized, toluene is a good solvent. It is used in synthetic resins, surface coatings and adhesives for this reason. It is also used as a start-

The structures of some of the simpler and better known aromatic hydrocarbons are illustrated below. As is apparent from the diagram, these substances consist of a single benzene ring with different functional groups attached — a methyl group ($-CH_3$) in toluene and a hydroxyl group ($-OH$) in phenol, for example.

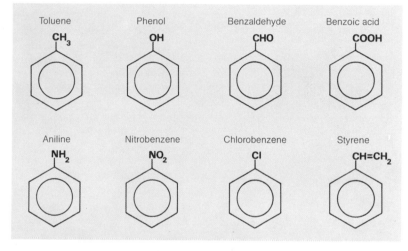

This unusual building (left) is a house made almost entirely of polyurethane foam, a polymer synthesized from a derivative of the aromatic hydrocarbon toluene ($C_6H_5CH_3$). To build the house, a thick layer of the foam was sprayed on the inside of large hemispherical molds, which were removed when the foam had hardened. The exterior surface was then painted with sunscreen paint to block ultraviolet light (which would otherwise cause the polyurethane to break down), and the interior surface was covered with a fire-proofing material (because polyurethane gives off poisonous fumes when burnt).

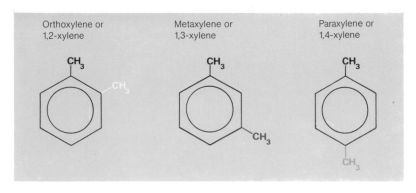

Orthoxylene or
1,2-xylene

Metaxylene or
1,3-xylene

Paraxylene or
1,4-xylene

Xylene has three isomeric forms (above). Orthoxylene has the second methyl group (—CH₃), shown in yellow, in the *ortho* position, adjacent to the primary methyl (in black). In metaxylene the second methyl (purple) is attached to the carbon atom one farther round the ring, in the *meta* position. And in paraxylene, the second methyl (blue) is in the *para* position, opposite the primary methyl group.

The adhesive under test below consists principally of the polymer polychloroprene (a synthetic rubber) dissolved in an aromatic solvent such as toluene.

ing material for the high explosive TNT (trinitrotoluene) and for assorted plastics, the most important being the urethane polymers. In addition, toluene is used in the manufacture of benzyl chloride, benzyl alcohol and benzaldehyde, and it is possible to convert it into a vast number of other derivatives. The major sources for toluene are coal tar and naphtha. It is extracted in the same way as benzene.

Xylenes

These are liquids at room temperature. They are similar to toluene in structure, except that a further hydrogen atom has been displaced from the ring by a methyl group (i.e. they are dimethylbenzenes). Three types of xylene exist — *ortho, meta* and *para* — depending on the relative positions of the two methyl groups about the ring. They are all useful solvents, and starting materials for synthesis of tri-substituted benzene derivatives. Paraxylene is used to make terephthalic acid, the starting material for the synthetic fiber Terylene (Dacron). Again, they can be prepared from tars

or naphtha using the same techniques as those for the isolation of benzene and toluene.

Condensed aromatic hydrocarbons

Benzene rings can be fused together to form a group of compounds called condensed-ring systems. The simplest of these is naphthalene, which appears to consist of two benzene rings joined together. Other common condensed-ring hydrocarbons include anthracene, which has three benzene rings fused together, and pyrene, which comprises four fused rings. These and many other far larger condensed hydrocarbons may be extracted in varying amounts from coal tar.

Naphthalene and anthracene

Best-known and simplest of the condensed-ring hydrocarbons, naphthalene is a solid at room temperature. Like anthracene — also a solid at room temperature — it can be isolated from coal tar.

Derivatives of both substances are used in the manufacture of dyestuffs. For instance, sulfuric acid converts naphthalene into naphthalene 1-sulfonic acid and naphthalene 2-sulfonic acid, intermediates for dyestuffs, and nitration gives 1-nitronaphthalene. This in turn can be converted into 1-aminonaphthalene, another potential precursor for the dyestuffs industry, but use of this compound has now been banned in many countries because of the presence of small amounts of a cancer-causing impurity (2-aminonaphthalene) during its formation. Naphthalene can also be oxidized to phthalic anhydride, an important starting material for the production of plasticizers and polyester resins.

Further condensed hydrocarbons can be made by fusing rings together. They can be added together to form a chain of rings, as in naphthacene, pentacene, hexacene, and so on, or joined honeycomb-fashion to give compounds such as phenanthrene, pyrene, benzapyrene, coronene and hexahelicene.

Coal carbonization is the major source of anthracene, naphthalene, and most other condensed aromatics. Coal is heated to above 570°F (300°C) in the absence of air. By the time the temperature reaches about 1,650°F (900°C), the coal has broken down into three main products — coal gas, coal tar, and coke. The coke is used in steel manufacture, and the gas as fuel gas. The other fraction, coal tar, contains a multitude of organic compounds, among them the condensed hydrocarbons. The most important ones, benzene, naphthalene and anthracene, are extracted using a combination of distillation, hydrogenation and liquid-liquid extraction techniques. Many other products are also contained in the coal tar, although in much smaller amounts.

Now that coal gas as a domestic heating fuel has given way in many places to natural gas, and less coke is required to smelt iron ore because blast furnaces are more efficient, coal carbonization is becoming less prevalent. Other sources of condensed hydrocarbons are available, however.

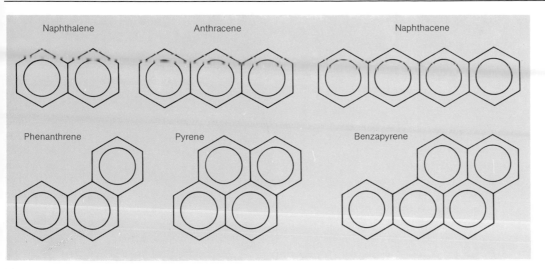

Condensed-ring aromatic compounds consist of two or more benzene rings joined together in chains of rings (as in naphthalene, anthracene and naphthacene; left, top row) or in a honeycomb-like arrangement (as in phenanthrene, pyrene and benzapyrene; left, bottom row). Although these structures are represented with all the carbon-carbon bonds being the same length — as in benzene — in fact the distances between adjacent carbon atoms are not equal.

These include steam reformates and pyrolysis gasoline.

Carcinogenic properties

All condensed-ring hydrocarbons are planar (except hexahelicene, which is helical) and all carbon-carbon bond lengths are similar in this group. But a more disturbing property they share is that of carcinogenicity (ability to cause cancer). Benzapyrene is the most dangerous in this respect.

Since these types of compounds are released when organic materials are heated to high temperatures, as in the carbonization of coal, for example, they have been linked to especially high incidence of skin cancers among those who have worked with coal tar.

Cigarette smokers are also thought to suffer the effects of condensed-ring compound toxicity, as these substances reside in the tar deposited in smokers' lungs. Scientists have now drawn positive links between heavy smoking and the incidence of lung cancer.

Graphite

Theoretically, sheets of condensed-ring hydrocarbons could stretch to infinity. Such compounds would be virtually identical with the large sheets of carbon rings that exist in graphite, a fairly common form of natural carbon. The fused rings are held together tightly, but the planes of atoms tend to slip over one another, making graphite a "slippery" substance. For this reason, it is sometimes used as a solid lubricant.

Tobacco smoke (left) contains hundreds of different chemical compounds, many of which are poisonous. Probably the most noxious of these substances are condensed-ring aromatic compounds, which are thought to be responsible for causing lung cancer because carcinogenicity is a property shared by all such compounds and because they have been found in high concentrations in tar in smokers' lungs.

The well known "pitch lake" at La Brea, Trinidad (above), is a naturally occurring deposit of asphalt, a bituminous substance that contains a large proportion of condensed-ring aromatic compounds. It is formed as a result of the volatile hydrocarbons evaporating from petroleum and the residue then being partly oxidized. Asphalt is used mainly for making road surfaces and as a waterproofing material

Halogenated hydrocarbons

It is comparatively easy to introduce halogens — chlorine, bromine, fluorine and iodine — into various organic compounds. This accounts for the vast number and range of organic halides that have been synthesized or exist in nature. Some have found specific uses, for instance as pesticides or starting materials for plastics, but organohalides are of most value because of their utility as chemical intermediates in a wide range of synthesis reactions.

Halogenation reactions

Usually, organic compounds can be halogenated by treatment with reagents such as hypohalous acids (for example, hypochlorous acid) or the halogen itself. Alkenes (olefins) are among the easiest compounds to halogenate. For example, ethene (ethylene, $H_2C=CH_2$) reacts with hypochlorous acid (HOCl) to form chlorohydrin ($ClCH_2$—CH_2OH), and with chlorine (Cl_2) to give 1,2-dichloroethane ($ClCH_2$—CH_2Cl).

Alkanes, cycloalkanes and aromatic compounds are also easy to halogenate; for example, bromine reacts with benzene to give bromobenzene.

Reactions of organic halogen compounds

Halogens can often be displaced from organic molecules by substituents that are more reactive, such as the hydroxide ion, ammonia and the hydrosulfide ion. An alkyl halide such as chloromethane (methyl chloride, CH_3Cl) thus forms methanol (CH_3OH), methylamine (CH_3NH_2) or methanethiol (CH_3SH). These types of conversions are called substitution reactions.

In elimination reactions, a complete molecule is displaced from the reagent being treated without anything taking its place. In this way alkyl halides yield alkenes and alkynes. For example, the elimination of hydrogen bromide (HBr) from 2-bromobutane ($CH_3CH_2CH(Br)CH_3$) yields butene ($CH_3CH=CHCH_3$); 2,2'-dibromopentane ($CH_3CH_2C(Br_2)CH_2CH_3$) gives 2-pentyne ($CH_3C=CCH_2CH_3$). Use of a strong base under vigorous conditions enables substitution of halogens in aromatic rings.

Another class of compounds important to organic chemists are the Grignard reagents, general formula RMgX, where R is an alkyl or aryl radical and X is a halogen. These are synthesized by treating an alkyl or aryl halide with magnesium:

$$RBr + Mg \rightarrow RMgBr$$

A Grignard reagent can be added to water to yield a hydrocarbon (RH):

$$RMgBR + H_2O \rightarrow RH + MgBrOH$$

Organolithium compounds are similarly useful and may be synthesized by adding lithium to organic halides. For example, with 1-chlorobutane:

$$CH_3CH_2CH_2CH_2Cl + Li \rightarrow$$
$$CH_3CH_2CH_2CH_2Li$$

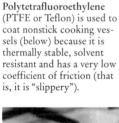

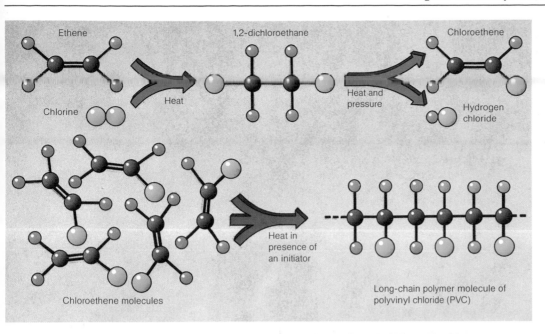

Ethene   1,2-dichloroethane   Chloroethene

Chlorine   Heat   Heat and pressure   Hydrogen chloride

Chloroethene molecules   Heat in presence of an initiator   Long-chain polymer molecule of polyvinyl chloride (PVC)

Polyvinyl chloride (PVC), a long-chain halogenated polymer, can be made from ethene and chlorine in three main stages (illustrated left). In the first stage ethene is heated with chlorine to produce 1,2-dichloroethane. In the second stage the 1,2-dichloroethane is heated under pressure to produce chloroethene and hydrogen chloride. After the hydrogen chloride has been removed, the pure chloroethene then remaining is heated in the presence of an initiator to form PVC.

Industrial uses of halogenated hydrocarbons

Few organic halides have direct use industrially, but are generally converted into other intermediates. 1,2-dichloroethane ($ClCH_2CH_2Cl$) is produced (from ethene) in the greatest quantities and used to make vinyl chloride ($CH_2=CHCl$), the building block for polyvinyl chloride (PVC).

Another important halogenated hydrocarbon is methyl chloroform (1,1,1-trichloroethane, $C(Cl)_3CH_3$), which is used for degreasing, cleaning drains, and as a solvent. It is made via several synthetic steps from vinyl chloride.

Natural gas provides the principal source of halogenated methane derivatives, which include trichloromethane (chloroform, $CHCl_3$) and tetrachloromethane (carbon tetrachloride, CCl_4). These two substances are used mainly as starting materials for the manufacture of fluorocarbon refrigerants called Freons. Examples are trichlorofluoromethane (Freon-11) and dichlorodifluoromethane (Freon-12). An important polymer, polytetrafluoroethylene (Teflon or PTFE), is derived from chlorodifluoromethane ($CHClF_2$), which on heating forms the monomer tetrafluoroethylene ($CF_2=CF_2$).

Another material produced from halogenated hydrocarbons is the synthetic rubber Neoprene. It is manufactured by polymerizing chloroprene ($CH_2=CCl-CH=CH_2$).

Effect on the ozone layer

In addition to their use as refrigerants, chlorofluorocarbons (CFCs) became employed as propellants in aerosol spray cans. But in 1974 it was claimed that discharge of CFCs into the atmosphere might deplete the ozone layer, which filters out harmful ultraviolet rays from the Sun, and this led to fears that if ultraviolet rays reached the Earth in greater amounts, it would cause increased incidence of skin cancer. Such was the outcry that CFCs were banned from use as propellants in the United States, although scientists have not yet been able to prove whether or not CFCs do deplete the ozone layer.

Chlorinated pesticides

Highly chlorinated hydrocarbons are among the most effective insecticides known. Most familiar of this group is DDT (2,2-di(p-chlorophenyl)-1,1,1-trichloroethane), which is termed a "hard" insecticide because it can exist in soil and water systems for months or even years. Other hard insecticides include Aldrin, Dieldrin and Lindane. Their use is severely restricted because although they successfully eradicate disease-transmitting insects and crop predators, they are also hazardous to humans. Many herbicides are also halogenated hydrocarbons. The most widely used is 2,4-D[(2,4-dichlorophenoxy) acetic acid], a wheat herbicide.

Records (above) are made largely of PVC, with the addition of plasticizers to make the records flexible. PVC is one of the best-known and most widely used of the vinyl polymers: in its rigid, unplasticized form it is used for pipes and guttering; and in its flexible, plasticized form for pliable sheeting, electrical insulation, non-inflammable upholstery coverings, floor tiles and clothing.

Alcohol (ethanol) has long been made by fermentation, in which enzymes from yeast cells (shown highly magnified, right) bring about the breakdown of starch or sugar into alcohol and carbon dioxide. The process is still the basis of the brewing industry for beer and lager and for the production of spirits such as whiskey (far right).

Alcohols

Alcohols are some of the most common and widely used chemicals. Methanol (methyl alcohol, CH_3OH) is the simplest, and has been used for many years as a solvent and as a fuel. Another well-known alcohol is ethanol, or ethyl alcohol (C_2H_5OH), which for thousands of years has been used as an intoxicating ingredient in alcoholic beverages such as wine, spirits, mead and ale. Ethanol is also important industrially as a solvent and as a reaction medium.

Structure of alcohols

All alcohols possess at least one hydroxyl (—OH) group in their molecular makeup. All simple alcohols are liquids. Besides their physical uses as solvents and reaction media, alcohols are of great value to the synthetic chemist because of their versatility in chemical reactions. They can be converted into almost every other kind of aliphatic compound, and form a whole host of important chemical intermediates. They can be converted easily into alkyl halides (RX, where R is an alkyl radical and X = F, Br, Cl or I), alkoxides (RONa, ROK, etc.), ethers (ROR'), esters (RCOOR') and alkenes (RC=CR'). Also, alcohols can be oxidized to form aldehydes, ketones and carboxylic acids, which all contain a carbonyl group (—C=O).

Methanol and ethanol

The simplest of all alcohols, methanol (CH_3OH) has a boiling point of 148.5°F (64.7°C). It was formerly known as wood alcohol because for centuries it was obtained by heating wood to about 900°F (about 500°C) in the absence of air and distilling off the liquid formed. Nowadays the chief source of methanol is synthesis gas, a mixture of carbon monoxide and hydrogen. This is passed at 572°F (300°C) and 300 atmospheres pressure over a catalyst:

$$CO + 2H_2 \rightarrow CH_3OH$$

Nearly half the methanol manufactured is converted into methanal (formaldehyde), a starting material for phenolic resins. Other derivatives include ethanoic acid (acetic acid), chloromethane (methyl chloride) and methyl methacrylate, the monomer for acrylic polymers.

With a boiling point of 172.9°F (78.3°C), ethanol (C_2H_5OH) is a colorless liquid at room temperature. Its preparation by fermentation is an ancient process used over many centuries for producing alcoholic beverages. Until the ascendency of the petrochemical industry, ethanol was prepared by fermenting plant-derived carbohydrates such as starch, sugar or cellulose. Today most industrial ethanol is prepared by hydrating ethene (ethylene, $H_2C=CH_2$) using a catalyst:

$$H_2C=CH_2 + H_2O \rightarrow C_2H_5OH$$

The major industrial use for ethanol (about 45 per cent of all uses) is in the production of ethanal

The boiling points of primary alcohols (red line) are higher than their isomeric secondary (purple line) or tertiary (orange) counterparts; isomers have the same molecular weight and lie on the same vertical line on the diagram. The low-molecular weight alcohols are miscible with water, due largely to hydrogen bonding between alcohols and water.

Boiling point

(°C)	(°F)	
	370	—CH₃
180		
		—CH₂
160		
		—CH
140		—C—
120		—OH
100	210	
	200	
80		
	170	
60		

Molecular weight: 30, 50, 70, 90, 110, 130, 150

Methanol, Ethanol, Propanol, 2-propanol, Butanol, 2-butanol, 2-methyl-2-propanol, Pentanol, 2-pentanol, 2-methyl-2-butanol, Hexanol, 2-hexanol, 2-methyl-2-pentanol, Heptanol, 2-heptanol, 2-methyl-2-hexanol, Octanol, 2-octanol, 2-methyl-2-heptanol

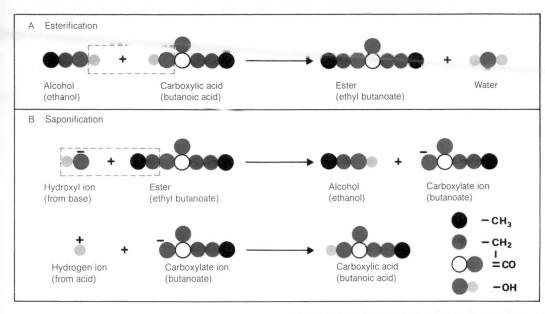

A Esterification

Alcohol
(ethanol) + Carboxylic acid
(butanoic acid) → Ester
(ethyl butanoate) + Water

B Saponification

Hydroxyl ion
(from base) + Ester
(ethyl butanoate) → Alcohol
(ethanol) + Carboxylate ion
(butanoate)

Hydrogen ion
(from acid) + Carboxylate ion
(butanoate) → Carboxylic acid
(butanoic acid)

$-CH_3$
$-CH_2$
$=CO$
$-OH$

Two key reactions of alcohols are esterification and saponification. Esterification (A) is the reaction between an alcohol and a carboxylic acid to form an ester and water; it is analogous to the neutralization of an acid by a base. Saponification (B) can be considered the reverse reaction. The action of a base on an ester followed by the addition of excess acid regenerates the original alcohol and acid.

(acetaldehyde, CH_3CHO), and about 30 per cent is used as solvent. The rest is used in various chemical processes.

Multi-functional alcohols

In addition to a single hydroxyl group, multi-functional alcohols contain other active substituents. These may be simply additional hydroxyl groups — as in the case of ethanediol (ethylene glycol, $HOCH_2—CH_2OH$) and glycerol — or other chemical groupings such as chloride or amino groups. Whatever the substituent, it can have considerable influence on the behavior and properties of the molecule.

The simplest of all multi-functional alcohols, ethanediol is a liquid at room temperature. It is an important ingredient of antifreeze and is manufactured industrially by hydrating oxirane (ethylene oxide, C_2H_4O).

Phenols

These are compounds based on phenol, a benzene ring linked to a hydroxyl group. Unlike most aliphatic alcohols, phenols are acidic. The hydrogen atom in the hydroxyl group can be detached as an acidic proton (H^+), because the benzene ring acts to stabilize the oxygen atom as an anion. Phenol is thus quite reactive and may be converted into a wide range of derivatives; about half the phenol produced is used to make phenolic resins, which are used in laminating, coatings and adhesives. A key industrial route to phenol is the acid-catalyzed rearrangement of cumyl hydroperoxide, which also yields dimethyl ketone (acetone).

Glycerol

A liquid at room temperature, glycerol or glycerine — 1,2,3-propanetriol — has many uses. It is employed as a moistener in the food, tobacco, cosmetic and drug industries, and also finds wide use as a lubricant, plasticizer and thickener. Nitroglycerin, the trinitrate ester of glycerol, is a well-known and powerful explosive. Glycerol itself can be synthesized in various ways or sapon-

ified from fats (which are esters of glycerol) in the manufacture of soap.

Epoxides

Although not strictly alcohols, epoxides are closely related — they possess an oxygen bridge between two carbon atoms. Oxirane (ethylene oxide), the simplest, is manufactured by oxidizing ethene (ethylene) and used in the industrial preparation of various compounds; about 50 per cent is used to make ethanediol (ethylene glycol). Another useful epoxide, 1,2-epoxypropane (propylene oxide), is utilized in the manufacture of polyester fibers, brake fluids and plasticizers.

Ethers

Ethers (general formula ROR') are formed when the hydroxyl hydrogen of an alcohol or phenol is replaced by another alkyl or aryl group. They are renowned for their inertness and for their utility as solvents. Best-known and most widely used of this group is ethoxyethane — ethyl, or "common", ether ($C_2H_5OC_2H_5$) — which has long been used as an anesthetic and as a solvent.

Like other polyhydric alcohols, ethanediol (ethylene glycol) has a high boiling point. This property is used in formulating antifreeze solutions for the radiators of motor vehicles that have to work at subfreezing temperatures.

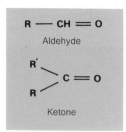

The general formulas of aldehydes and ketones show that they both have a carbonyl (C=O) group. In an aldehyde, it is usually joined to a hydrogen atom and an alkyl or aryl group (R). In a ketone, the carbonyl is bonded to two such groups (R and R').

Aldehydes and ketones

Both aldehydes and ketones are classes of organic compound that contain carbonyl groups — that is groups in which a carbon atom is doubly bonded to an oxygen atom (—C=O). In aldehydes, the carbonyl group is linked to two hydrogen atoms, or to one hydrogen and an alkyl or aryl group. Methanal (formaldehyde, HCHO) is the simplest and most familiar. In all other aldehydes, the carbonyl group bears just one hydrogen atom, the other portion of the molecule being an alkyl or aryl group. An example is ethanal (acetaldehyde, CH_3CHO), which is identical to methanal except that one of the hydrogen atoms has been replaced by a methyl group.

Ketones are different structurally, in that the carbonyl group bears no hydrogen atoms. Instead, it is linked to two alkyl or aryl groups. Propanone (also called dimethyl ketone or acetone, $(CH_3)_2CO$), the most widely used and familiar ketone, consists of a carbonyl group bonded to two methyl groups.

Preparation and uses

Aldehydes are usually made by oxidizing a primary alcohol (with potassium dichromate):

$$RCH_2OH \rightarrow RCHO \text{ (aldehyde)}$$

and ketones can be prepared by oxidizing a secondary alcohol (with potassium permanganate):

$$RCH(OH)R' \rightarrow R.CO.R' \text{ (ketone)}$$

Aldehydes and ketones have a variety of uses, although most are employed as solvents or as chemical reagents. Both types of compounds can be converted into alcohols and alkanes by successive reduction reactions. Moreover aldehydes can be oxidized to form carboxylic acids.

Low-molecular weight aldehydes have unpleasant pungent smells, whereas the odors of ketones, widely distributed in nature, are sometimes agreeable enough to warrant their use as flavorings or perfume ingredients. Examples are biacetyl, a principal flavoring component of margarine, and muscone, an expensive perfume ingredient. Camphor, a ketone, has long been valued for its medicinal properties and testosterone, also a ketone, is the sex hormone that promotes development of male characteristics in humans and other mammals.

Methanal and other simple aldehydes

With a boiling point of −2°F (−19°C), methanal is the only member of the aldehyde family which is gaseous at room temperature, although ethanal has a boiling point of just 70°F (21°C). All the other low-molecular weight aldehydes are liquids at room temperature. When passed as a gas through glass apparatus, methanal spontaneously forms a polymer called paraformaldehyde. It appears as a white coating on the glass, and if heated to above 400°F (above 200°C) is reconverted into gaseous methanal.

A well-established use for methanal is as a preservative fluid for biological specimens. It is first dissolved in water to form a 37 per cent aqueous solution called formalin, which is also important in the manufacture of resins and plastics. Ethanal is also capable of forming polymers. Under acidic conditions, the trimer paraldehyde is formed. This was used medicinally as a sedative and hypnotic, but has dangerous side effects which have made it obsolete. Under slightly different conditions, ethanal forms the tetramer, metaldehyde, which can be used as fuel or as a slug poison.

Propanone and simple ketones

Like all other simple ketones, propanone is a liquid at room temperature; it has a boiling point of 133°F (56°C). It mixes well with water and like butanone (methyl ethyl ketone), another simple ketone, is an excellent solvent for organic reactions. Although the carbonyl group in ketones is less reactive than that in aldehydes, it is still prone to attack by other chemicals. Like aldehydes, ketones can be converted into alcohols by the addition of hydrogen in the presence of suitable catalysts, although higher pressures of hydrogen gas are needed to hydrogenate ketones. Acetals, imines and cyanohydrins are other useful chemicals formed from ketones and aldehydes.

Acetals

When mixed with an excess of alcohol, aldehydes react to form acetals and hemiacetals, which are useful in chemical synthesis. During the reaction, the double bond in the carbonyl group is destroyed by an attacking alcohol molecule. This adds on to the aldehyde molecule to form a hemiacetal, which has one ether grouping and one hydroxyl group in place of the carbonyl group. Acetals are formed when hemiacetals are subjected to further attack by alcohol, and bear two, instead of just one, ether groupings.

Aldehydes are easily oxidized, and can therefore act as reducing agents. This fact is used in detecting sugar (in the form of glucose, which has an aldehyde grouping) in urine samples in testing for the presence of diabetes mellitus, a hormone deficiency disorder in which the patient's body cannot properly metabolize glucose. The color changes are caused by the precipitation of finely divided copper (I) oxide from Fehling's solution, which is a complex formed from copper (II) sulfate and potassium sodium tartate (Rochelle salt) in alkaline solution.

Urine sample

Fehling's solution

Urine aded to Fehling's solution and mixture warmed

No sugar (normal)

Slight sugar (mild diabetic)

High sugar (diabetic)

Formalin — an aqueous solution of ethanal (formaldehyde) — has long been used for preserving biological specimens. For this purpose it is usually preferred to ethanol (alcohol), which dehydrates specimens and distorts their tissues. The pungent smell and irritating fumes of formalin are not, however, liked by biology students.

The carbonyl group in ketones may be similarly modified to yield hemiketals and ketals. As acetals and ketals broadly share the same chemical properties, they are usually referred to collectively as acetals.

Imines are formed by reacting aldehydes or ketones with amines. Some of them are important biological intermediates. Their formation is a key step in the metabolism of amino acids, the building blocks of proteins. Chemists use cyanohydrins (made by reactions of aldehydes and ketones with hydrogen cyanide) in chemical synthesis.

Keto-enol tautomerism

In chemical synthesis, a useful feature of ketones is that they can exist in two forms, depending on the acidity of their surroundings. Interconversion between the two forms is known as tautomerism, and results from the shuttling of a hydrogen atom between the oxygen on the carbonyl group and the carbon of a neighboring alkyl group. In the keto form, the carbonyl group remains intact. But once a suitable base or acid is added, a proton (hydrogen atom) can be induced to migrate from a neighboring alkyl group to the oxygen atom on the carbonyl group. The net effect is to transform the ketone into a vinyl alcohol, possessing a carbon-carbon double bond and a hydroxyl grouping. This is known as the enol form. Synthetic chemists can easily tip the equilibrium either way depending on which form (or tautomer) they require for use. Under ordinary conditions simple ketones contain only a fraction of the enol form. The opposite is true of 1,3-dicarbonyl compounds, in which two carbonyl groups are separated by a methylene group. In these, the enol form may total as much as 80 per cent.

Ketones, particularly propanone (acetone) and cyclohexanone, find major uses as industrial solvents, especially for cellulose compounds such as those used in paints and lacquers. The formulation called cellulose thinner is used to make cellulose paint thin enough for spraying.

A simple ketone such as propanone (acetone) exists almost entirely as the keto form (A) with its oxygen atom doubly bonded to the carbon atom of the carbonyl group. A tiny proportion changes reversibly to the enol form (B) in which the oxygen atom is joined to a hydrogen atom. This reversibility is called tautomerism. The more complicated diketone 2,4-pentadione (C), with two carbonyl groups, exists mainly as the enol form (D) because of the extra stability conferred by its six-membered ring.

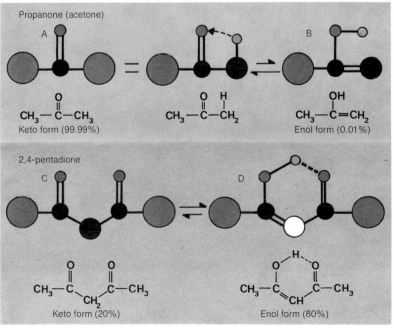

Propanone (acetone)

A

$$CH_3 - \overset{\overset{O}{\|}}{C} - CH_3$$
Keto form (99.99%)

$$CH_3 - \overset{\overset{O}{\|}}{C} - \overset{\overset{H}{|}}{CH_2}$$

B

$$CH_3 - \overset{\overset{OH}{|}}{C} = CH_2$$
Enol form (0.01%)

2,4-pentadione

C

$$CH_3 - \overset{\overset{O}{\|}}{C} - \underset{\underset{CH_2}{}}{} - \overset{\overset{O}{\|}}{C} - CH_3$$
Keto form (20%)

D

$$CH_3 - \overset{\overset{O \cdots H \cdots O}{}}{C} = CH - \overset{}{C} - CH_3$$
Enol form (80%)

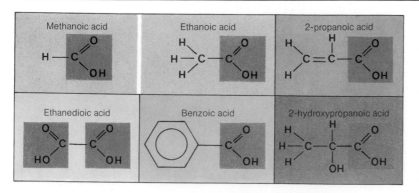

Methanoic acid | Ethanoic acid | 2-propanoic acid
Ethanedioic acid | Benzoic acid | 2-hydroxypropanoic acid

There are various types of organic acids but all contain at least one carboxyl group (—COOH). The diagram above illustrates the main types: saturated aliphatic, such as methanoic and ethanoic acid (blue box); unsaturated aliphatic, such as 2-propanoic acid (purple box); dicarboxylic, such as ethanedioic acid (pale green box); aromatic, such as benzoic acid (darker green box); and polyfunctional (containing one or more functional groups in addition to the carboxyl group), such as 2-hydroxypropanoic acid (pale brown box).

The venom of ants (seen attacking a wasp, below left) and the bark of willow trees (below right) both contain organic acids — methanoic acid in the ants' venom and salicylic acid in willow bark.

Organic acids

Organic acids are a group of compounds characterized by the carboxyl group (written as —COOH or —CO$_2$H), which is itself made up of a carbonyl group (—C=O) and a hydroxyl group (—OH). These compounds are commonly called carboxylic acids.

Organic acids are weak compared to mineral acids. This is because, when in solution, they dissociate less completely and so produce a lower concentration of hydrogen ions than do mineral acids. The most common type of saturated, aliphatic carboxylic acid has the general formula R-COOH, where R is an alkyl group. But some acids are unsaturated, or contain a ring system, or have two carboxyl groups (a dicarboxylic acid).

Organic acids are named by dropping the final *e* from the name of the longest hydrocarbon chain containing the carboxyl group (including the carboxyl carbon) and adding —*oic acid.* For example, C$_2$H$_5$COOH is propanoic acid, and CH$_3$COOH (still commonly called acetic acid) is properly termed ethanoic acid.

Physical properties and structure
Physical properties depend on the substituents, but typical simple aliphatic acids are pungent,

corrosive, water-soluble liquids. High molecular weight saturated acids are generally odorless, waxlike solids. Boiling points of lower acids are higher than expected because of dimerization (two molecules are linked together by relatively weak hydrogen bonds). The structural geometry of carboxylic acids requires that the C=O bond in the —COOH group be shorter than the normal C=O bond, and this is confirmed by measurements on crystalline acids. But the C—OH bond in acids is always shorter than expected as well. Studies of the carboxyl anion (—COO$^-$) show that both carbon-oxygen bonds are in fact the same length. This is because the structure is said to be resonating between two equivalent forms. The actual structure has the negative charge delocalized (spread between several centers — in this case the COO group) to form a stable hybrid, where each carbon-oxygen bond has the same degree of multiple-bond character. Some of this resonance is always occurring in the acid, and this is why the C—O bonds are always shorter than expected.

Chemical properties
Carboxylic acids can react to form salts, esters, anhydrides, acid halides and amides or be reduced to form aldehydes and alcohols. Soaps are alkali metal salts of fatty acids (aliphatic acids made up of long alkyl chains terminating in a carboxyl group), made by treating oils or fats (esters) with alkali metal hydroxides (such as sodium hydroxide).

In aqueous solution, the soap molecule has a hydrocarbon end (which is hydrophobic — "water fearing") and a charged carboxyl end (which is hydrophilic — "water loving"). Soaps work by surrounding dirt or grease particles with many soap molecules; the hydrocarbon ends dissolve in the grease or dirt particle and the ionized carboxyl ends point outwards, dissolving in the water. Thus the soap molecules form an emulsion — a permanent suspension or dispersion, usually

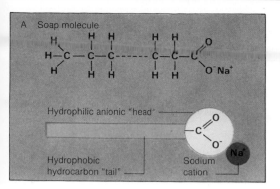

A Soap molecule

Hydrophilic anionic "head"

Hydrophobic hydrocarbon "tail"

Sodium cation

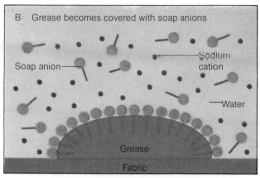

B Grease becomes covered with soap anions

Soap anion

Sodium cation

Water

Grease

Fabric

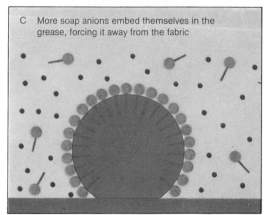

C More soap anions embed themselves in the grease, forcing it away from the fabric

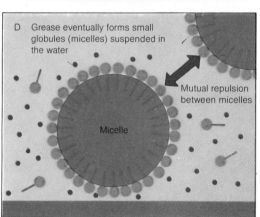

D Grease eventually forms small globules (micelles) suspended in the water

Mutual repulsion between micelles

Micelle

Soaps are alkali-metal salts of long-chain aliphatic acids, usually the sodium (in hard soaps) or potassium (soft soaps) salts of oleic acid, $CH_3(CH_2)_7$ $CH=CH(CH_2)_7COOH$, palmitic acid, $CH_3(CH_2)_{14}COOH$, or stearic acid, $CH_3(CH_2)_{16}COOH$. Diagram A (far left) illustrates the structure of a typical hard soap (top) and a simplified representation (bottom). Diagrams B to D (near left) show the detergent action of soap. In aqueous solution the sodium cation separates from the main part of the molecule, the hydrophobic tail of which embeds itself in grease (B). The grease is gradually forced away from the fabric as more soap molecules embed themselves (C). Eventually the grease separates from the fabric, forming a micelle (D). Micelles repel each other and so remain suspended in the water.

During vigorous activity (below) lactic acid accumulates in muscles as a result of incomplete oxidation of glycogen (the energy source), itself due to a lack of oxygen. When activity stops, some lactic acid is oxidized (to carbon dioxide and water) and some is converted back to glycogen.

of oil or fat in water — with the dirt or grease, enclosing it in a water-solubilizing envelope (called a micelle). In other words, soap molecules package dirt in droplets which can be taken into solution and washed away.

Carboxylic acids are important in nature (all amino acids are carboxylic). A number of higher molecular weight acids are isolated from animal and vegetable sources (as triglycerides — esters of glycerol, the trialcohol, 1,2,3-propanetriol) for industry. Lower acids can also be prepared synthetically. Methanoic (formic) acid (HCOOH) is the simplest monocarboxylic acid and it occurs in the venom of bees and ants. Ethanoic (acetic) acid (CH_3COOH) is the end product of fermentation of carbohydrates by some organisms. It is also the active ingredient in vinegar. Propanoic (propionic) acid (C_2H_5COOH), the next member in the series of saturated aliphatic carboxylic acids, is found in milk, butter and cheese; it is the first carboxylic acid (that is, the one with the lowest molecular weight) to exhibit some of the properties of fatty acids.

Ethanedioic (oxalic) acid is the simplest dicarboxylic acid (HOOC—COOH) and it is formed by oxidation of carbohydrates. It occurs naturally in many plants, notably rhubarb. Many other fatty acids (with between 14 and 22 carbon atoms) are oxidized in animals and plants to provide energy.

Examples of aromatic carboxylic acids are benzoic acid and salicylic acid, both of which occur naturally. Benzoic acid appears in the juices of berries (especially cranberry) and it is used as a preservative in many foods. Salicylic acid (found in willow bark) reacts with acetic anhydride, a derivative of acetic acid, to give acetylsalicylic acid, otherwise known as aspirin.

Polyfunctional acids include tartaric acid, found in grapes and important in the discovery of optical isomerism by Louis Pasteur (1822—1895), and lactic acid, responsible for the acidity of sour milk and also formed in muscle tissue during exercise.

Citrus fruits owe their sharp flavor to citric acid, whereas malic acid gives apples their tartness. All of these acids are key biochemical intermediates in the metabolism of sugars and fats. Citric, acetic, tartaric and propionic acids are commonly used as food additives and are known as acidulants. They are used to prevent rancidity, enhance flavors and modify texture in foods.

Esters

Esters are organic compounds that result from a condensation reaction between an acid (usually a carboxylic acid) and an alcohol. The reaction is called esterification and it was originally thought to be analogous to neutralization, the reaction between an inorganic acid and a base. But the mechanism is different. Nevertheless, it is an important chemical reaction in organic synthesis.

The general formula for an organic ester is RCOOR', where R and R' are alkyl or aryl radicals. An ester is named after the alcohol and acid from which it is derived. Ethanoic (acetic) acid gives ethanoates (acetates) such as ethyl ethanoate, whereas butanoic (butyric) acid gives butanoates (butyrates). Animal fats, which are esters of carboxylic acids with the polyfunctional alcohol glycerol (1,2,3-propanetriol), are called triglycerides.

Esters are moderately polar, but are generally insoluble in water and have boiling points slightly higher than hydrocarbons of similar molecular weights. Low molecular weight esters are usually colorless liquids and high molecular weight esters are generally solids. They frequently have pleasant odors and are responsible for the fragrance and flavor of many flowers and fruits. Cyclic esters are known as lactones.

Formation of esters

Esterification generally takes place in a nonaqueous solvent, with an acid as a catalyst, to give an ester and water:

$$R'COOH \text{ (acid)} + ROH \text{ (alcohol)} \rightarrow R'COOR + H_2O$$

However, the reaction can be between an acid anhydride or acid chloride and an alcohol, or between an acid salt and an alkyl halide. Trans-esterification takes place between an ester and an alcohol to form a new ester.

Although an organic ester may be said to correspond to a salt in inorganic chemistry, it is formed in a different way. Whereas the hydrogen from the mineral acid and the hydroxyl group from the base form water, leaving a salt, it is generally the hydrogen from the alcohol and the hydroxyl group from the organic acid which form water, leaving an ester.

Esterification has been studied extensively by physical chemists since the middle of the nineteenth century, because it involves an equilibrium — in other words, it is a reversible reaction. Hydrolysis is the reaction which splits the ester to give the parent acid and alcohol. It is usually catalyzed by dilute mineral acid or a base, in the presence of water, at high temperature to speed it up.

The irreversible reaction of an ester with a base forms an alcohol and the salt of the acid. This type

A

Banana flavor

$$CH_3COOCH_2CH_2CH(CH_3)_2$$

Isoamyl ethanoate

B

Beeswax

$$CH_3(CH_2)_{14}COO(CH_2)_{29}CH_3$$

Myricyl palmitate

C

Tallow (candle wax)

$$CH_2OCO(CH_2)_{16}CH_3$$
$$CH\ OCO(CH_2)_{16}CH_3$$
$$CH_2OCO(CH_2)_{16}CH_3$$

Glyceryl tristearate

D

Aspirin

COOH

OCO CH₃

2-(Acetyloxy)benzoic acid

E

Lucite
polymer of

$$CH_3$$
$$CH_2 = C\ COOCH_3$$

Methyl methacrylate

Esters occur widely in nature and have many practical uses. Some give the characteristic smell and flavor to fruits, such as isoamyl ethanoate (A) in bananas. Many waxes and oils consist mainly of esters, such as myricyl palmitate (B) in beeswax and glyceryl tristearate (C) in tallow. The painkilling drug aspirin is an ester-hydroxyl-benzoic acid (D) and the plastic Lucite, used for making toothbrush handles and spectacle frames, is a polymer of the ester methyl methacrylate (E). Coconut oil, used in soapmaking and obtained from the fruit of the coconut palm tree (far right), contains various fatty acid esters.

The subtle flavors of wines depend largely on the presence of small amounts of esters. They continue to form gradually after fermentation has ceased, while the wine "ages" in the racked bottles stored in a cool cellar.

of hydrolysis is historically called saponification and it is one of the oldest chemical reactions known. Soap was made by saponification of natural fats and oils (lard or tallow) with lye (aqueous sodium or potassium hydroxide).

The mechanism of hydrolysis has been studied carefully to see which carbon-oxygen bond breaks in the ester. Using water which contains a rare isotope of oxygen (^{18}O) it is possible to find out whether the ^{18}OH group adds to the acid or the alcohol and so determine which bond breaks. Except in a few cases, only the acid contains the ^{18}O, and so it must be the acid carbon-oxygen bond which breaks.

Uses of esters

Low molecular weight esters (such as ethyl ethanoate and butyl ethanoate) are used industrially as solvents for lacquers, resins and varnish. High molecular weight esters may be used as softeners for plastics. The plastic polymethyl methacrylate, a glass substitute known as Lucite or Plexiglas, is made using the ester of an unsaturated acid. Polyethylene terephthalate, the basis of textile fibers such as Terylene, Fortrel and Dacron and used in sheets as photographic film (Mylar), is made by polyesterification. Aspirin is an ester — the product of the reaction between salicylic acid and acetic anhydride. Cellulose esters are used in photographic films, as a textile fiber (rayon), and in explosives. Natural oils and waxes are largely simple esters; beeswax is mainly myricyl palmitate.

Low molecular weight esters have characteristic fruitlike odors. The flavor and odors of many fruits and flowers are due to a complex mixture of minute traces of esters. Thus esters have been widely used as synthetic flavors and in perfumes; for example, the taste and smell of bananas (isoamyl ethanoate), rum (isobutyl propanoate), pineapple (butyl butanoate), apple (isoamyl val-

erate) and wine/cognac (ethyl decanoate). Wines have hundreds of esters in trace amounts contributing to their flavor. The ethyl esters of the wine acids form slowly, adding to the unique bouquet of vintage wines, rum and brandy.

Lactones are usually encountered in their most stable form — the five- and six-membered rings respectively known as gamma- and delta-lactones. Both types are widely distributed in nature, especially in plants. Large ring lactones (with 15 or 16 carbon atoms) have musky odors. Several of the unsaturated lactones found in plants (for example, the angelica lactones) are used as drugs.

Many old aircraft, and some modern lightweight ones, have surfaces made by fixing fabric (such as linen) onto wooden frames. A key stage involves painting the fabric with a quick-drying lacquer or "dope" consisting of, typically, a cellulose compound dissolved in a volatile ester, such as amyl ethanoate.

Nitrogen compounds

There is a wide range of organic compounds that contain nitrogen, including amines, amides, nitriles, oximes and nitro compounds. Among the most important are amino acids, and these have an article later in this book.

Amines can be considered to be derivatives of ammonia (NH_3). They are classed as either primary, secondary or tertiary, according to the number of hydrogen atoms "replaced" by organic groups. If one hydrogen is replaced, the compound is a primary amine (general formula RNH_2); if two hydrogens are replaced by organic groups, it is a secondary amine ($RR'NH$). If three organic groups are attached to the nitrogen atom, it is a tertiary amine and has no hydrogen atoms connected directly to the nitrogen atom ($RR'R''N$).

Aliphatic amines are named after the parent hydrocarbon. Thus, if one of the hydrocarbon atoms in methane (CH_4) is replaced by an amino ($-NH_2$) group, the resultant compound is aminomethane (formerly called methylamine, CH_3NH_2). The more complicated secondary and tertiary amines are named as *N*-substituted derivatives of a parent primary amine. Thus *N*-methylaminomethane is the secondary amine $(CH_3)_2NH$. Many aromatic amines (such as aniline) and nitrogen ring compounds are still known by their trivial names, because their proper names are unwieldy.

Amines play important roles in biochemical systems and are widely distributed in nature as amino acids, alkaloids and vitamins. Manufactured amine derivatives are medicinal chemicals (sulfa drugs and anesthetics) and precursors of synthetic fibers such as nylon. Aniline (aminobenzene) is the most important industrial amine; it is highly toxic and it can be used to make many substituted benzene derivatives, including a whole family of dyestuffs.

Structure and properties of amines

The nitrogen atom in an amine has a lone pair of electrons and it can be considered as a base, like ammonia. Amines with less than five carbon atoms are miscible with water. Aqueous solutions of amines are basic and almost all amines are soluble in dilute acid because they form soluble ammonium salts.

Primary and secondary amines do not form intermolecular hydrogen bonds as strongly as do alcohols, so their boiling points lie between those of the corresponding hydrocarbons and alcohols. Tertiary amines have boiling points slightly higher than the corresponding hydrocarbon. Thus the methylamines are all gases at room temperature. Simple amines have a distinct odor of rotting fish, whose characteristic smell is caused by bacterially-produced amines.

Preparation and reactions

Formation of primary amines is most important because secondary and tertiary compounds can be made from them. Synthesis is by treating an alkyl halide with ammonia or sodium azide and hydrogen, or by the reduction of nitrogen-oxygen compounds (such as amides, oxides, nitro compounds or nitriles). The Hofmann degradation and the Curtius rearrangement are important synthetic steps that can be used to form amines from acyl (CO-group) compounds with one more carbon atom.

Primary amines can be converted to secondary amines by alkylation, to secondary amides by acylation (an amide is an ammonia or amine derivative in which the nitrogen atom is attached to the organic group through an acyl linkage) and to imines (organic compounds with a carbon-nitrogen double bond). Imines are very reactive and can be considered as the nitrogen analogue of a carbonyl compound. Secondary amines react

Ammonia

Primary amine

Secondary amine

Tertiary amine

Ammonia (NH_3), a simple nitrogen compound, is the "parent" of amines. In a primary amine, one of ammonia's hydrogen atoms is replaced by an alkyl or aryl group (R). Substitution by further groups gives secondary and tertiary amines.

Amines, and proteins derived from them, occur in the tissues of all animals. In dead fish, particularly, amines reveal their presence through their characteristic "fishy" smell.

similarly but less vigorously, except that they form enamines (analogous to an enol, where the nitrogen is connected to an olefinic carbon) instead of imines. Tertiary amines can be alkylated to give quaternary ammonium ions, equivalent to an ammonium ion (NH_4^+) with each of the hydrogen atoms replced by an alkyl group.

Heterocyclic nitrogen compounds

Heterocyclic compounds are ring structures that include at least one atom other than carbon in the ring. Aromatic heterocyclic compounds of nitrogen (containing a nitrogen atom in the ring) are important as biochemical intermediates. In some respects their chemical properties are similar to benzene derivatives — for example, they may display aromatic character. One of the most common is pyridine, a six-membered ring with one carbon substituted by a nitrogen. It reacts as amines do with acids or alkyl halides to give pyridinium salts. It can be extracted from coal tar and its derivatives are important in nature (for example, nicotinic acid).

Quinoline and isoquinoline are analogues of naphthalene, being a fusion of benzene and pyridine. Their reactions are similar to those of each constituent and a small number of derivatives occur naturally in plants as alkaloids. For example, quinine occurs in the bark of the cinchona tree and is used as an antimalarial agent and as tonic-water flavoring.

Pyrrole is a five-membered heterocycle (with four carbons and one nitrogen in the ring), considered to be aromatic, which can be isolated from coal tar or made industrially from ammonia and furan. Substituted pyrroles can be prepared easily by a variety of synthetic methods.

An important derivative of pyrrole is indole, formed when a benzene ring is fused to a pyrrole ring. It can be synthesized by the Fischer indole synthesis, which involves heating a phenylhydrazone of an aldehyde or ketone under strongly acidic conditions. Indole crystals are colorless

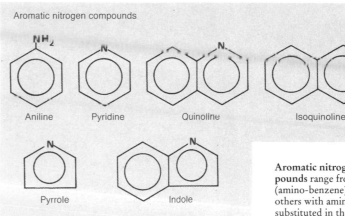

Aromatic nitrogen compounds

Aniline   Pyridine   Quinoline   Isoquinoline

Pyrrole   Indole

Aromatic nitrogen compounds range from aniline (amino-benzene), and others with amino groups substituted in the ring of benzene or one of its analogues, to heterocyclic compounds in which a nitrogen atom has a place in the ring itself.

with a pleasant odor, and it can be used in perfumes. Indole undergoes substitution reactions more readily than benzene. Its derivatives are widespread in nature; many alkaloids and hallucinogens are derived from tryptophan, an amino-acid derivative of indole.

Although few pyrroles occur in nature, the porphyrins are a class of naturally occurring pigments which can be said to have originated from four pyrrole units linked with C—H bridges. They are not actually pyrrole derivatives but a separate class of highly-colored stable aromatics. The porphyrin system is of great importance because it forms what are termed chelate compounds with metal ions, holding the metal between the four nitrogens. This is the basis for the red blood pigment hemoglobin and the green leaf pigment chlorophyll. Vertebrate blood is able to carry oxygen because of hemoglobin — the key is a six coordinate iron (II) atom, held on four sides by the nitrogens in heme (the porphyrin complex), and held underneath by a nitrogen ligand from the protein globin. The sixth ligand is oxygen (O_2), bound reversibly because the long protein chain prevents other suitable ligands from approaching close enough to join the oxygen.

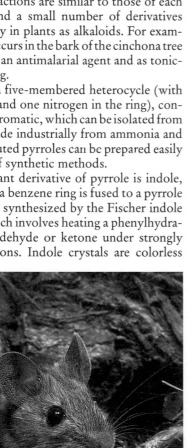

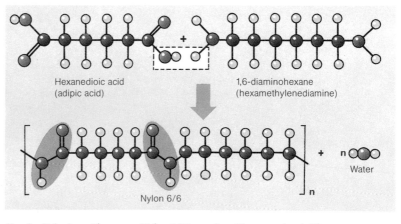

Hexanedioic acid (adipic acid)   1,6-diaminohexane (hexamethylenediamine)

Nylon 6/6   +   n   Water

Simple aliphatic amides, containing the —CONH— group, can also be recognized by their distinctive smell, which is responsible for the odor of mice (left).

Nylon 6/6 is a polyamide polymer, built up by the formation of amide linkages between an acid and an amine, with the elimination of water (see diagram

above). The reactants are heated to a temperature of 518°F (270°C) under a pressure of 10 atmospheres to make the reaction take place.

Warning eye-spots on a South American butterfly's wing can be seen in close-up as a series of overlapping scales. The pigments in the colors belong to a class of substances called pterins, which are compounds with a double-ring system containing four nitrogen atoms and six carbon atoms.

group) and hydroxylamine (NH_2OH). Oximes are characterized by the C=NOH group. They may be aldoximes (derived from aldehydes) or ketoximes (derived from ketones), but both have a carbon-nitrogen double bond — there is restricted rotation about the bond axis and therefore geometrical isomers can be isolated. The resulting compounds are termed *syn* and *anti;* these prefixes refer to the position of the hydroxyl group relative to the other groups. However, few pairs are actually isolated, probably because conversion from one isomer of the C=N double bond to the other is relatively easy compared to a carbon-carbon bond system.

Oximes are used to isolate and identify carbonyl compounds; some are important industrial chemicals. Aldoximes may be dehydrated to nitriles by using acetic anhydride, and ketoximes may be reduced to primary amines. Oximes may be hydrolyzed to carbonyl compounds and hydroxylamine salts by heating with an aqueous mineral acid. This hydrolysis is sometimes difficult because oxime formation is reversible, so formaldehyde is added because it combines with free hydroxylamine to give a very stable oxime. Ketoximes may become amides by the Beckmann rearrangement; an example of this important reaction is in the formation of caprolactam, an intermediate in the manufacture of Nylon 6.

Nitro compounds

Organic nitro compounds are derived from hydrocarbons, having NO_2 groups with nitrogen-carbon bonding. Nitration, the formative reaction, is usually an exothermic reaction and the nitro group tends to hinder further nitration. Aliphatic nitro compounds, the nitroparaffins, are polar, relatively acidic and can be used as solvents. They are also important in organic synthesis.

Aromatic nitro compounds, known for more than a century, are chiefly used as dye intermediates and as explosives. They are formed readily by the reaction of aromatic hydrocarbons and nitric acid. Examples are trinitrotoluene (TNT) and picric acid (trinitrophenol). The most important synthetic reaction of the aromatic nitro compounds is reduction to aromatic amines. As the simplest and most practical synthesis of such compounds, it has been extensively studied and many reagents are suitable; for example, hydrogen with a catalyst (Pd, Ni, Pt) or iron, tin or zinc with a mineral acid.

Nitrobenzene is an important industrial product, used as a solvent and to synthesize aniline. Nitrotoluene is also an important derivative — it is used in the manufacture of trinitrotoluene (TNT). The process is carried out in three steps; first the toluene is added to the acid at 284°F (140°C) to produce mononitrotoluene; this is separated and then trinitrated at high temperature. The product is recrystallized carefully and prepared for use.

Nitroglycerin (really a nitrate ester of glycerol) is another high explosive, although not a true nitro compound and not an aromatic. It was con-

Chlorophyll, the green pigment in plants, is essential in photosynthesis. It is a dihydroporphyrin magnesium complex similar to heme — suggesting a common evolutionary background for animals and plants.

Organic cyanides

Nitriles (general formula RCN) are a series of organic cyanides. They are the only simple functional nitrogen compounds whose synthesis involves the formation of a carbon-carbon bond.

Nitriles are weak bases and are highly poisonous. They are important in synthetic chemistry because they can be hydrolyzed to give carboxylic acids or reduced to amines. They thus provide the synthetic chemist with a relatively simple means of adding a new carbon atom to a molecule. Acrylonitrile is an important starting material in the manufacture of fabrics, plastics and rubbers.

Nitrogen-oxygen compounds

An important branch of organic chemistry is concerned with compounds with functional groups containing nitrogen and oxygen. These may be nitro compounds, nitroso compounds, or oximes. Organic nitro compounds are formed by nitration, the means by which a nitro (NO_2) group is added to a hydrocarbon, generally by using nitric acid. Nitroso compounds are those which contain an NO group. Important related compounds are the nitrosamines.

An oxime is the result of a condensation reaction between an aldehyde or a ketone (carbonyl

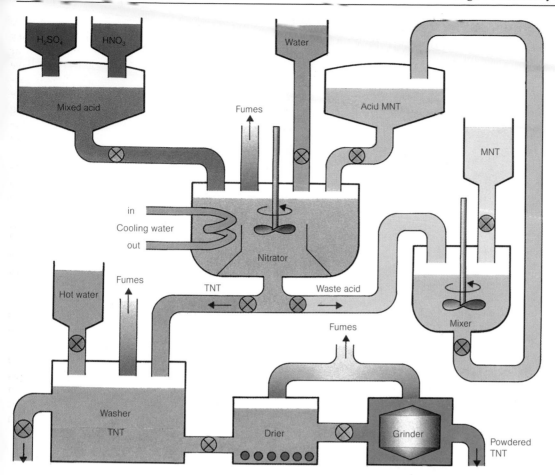

Many explosives, including detonators, propellants and warheads, are organic nitrogen compounds. One of the major high explosives for military use is TNT (trinitrotoluene). The diagram (left) shows an industrial process for the continuous production of TNT from MNT (mono-nitrotoluene), which is itself made by nitrating toluene ($C_6H_5.CH_3$) using a mixture of sulfuric acid (H_2SO_4) and nitric acid (HNO_3).

Modern missiles, such as Seawolf (below), developed as a naval antimissile missile, usually have warheads containing nitro-explosives. Many military explosives are mixtures — often containing TNT — chosen to be insensitive to heat and shock except when deliberately detonated.

sidered too dangerous to use until Alfred Nobel (1833—1896) found a way to stabilize it, absorbing it in a claylike mineral to make dynamite.

Nitroso compounds

Compounds with the nitroso group are rather rare. This is because many primary and secondary nitroso compounds readily isomerize in an acid or base to give oximes. Tertiary derivatives have a bluish color in the gas phase and in dilute aqueous solution — they are colorless liquids in the pure state.

Nitrosation is the reaction of amines with nitrous acid. It is a complex reaction usually carried out by adding aqueous sodium nitrite ($NaNO_2$) to a cold aqueous solution of the amine in dilute mineral acid. The reactive nitrous acid is formed in solution. In aromatic amines, the nitroso group generally attaches to the ring-carbon atom. With aliphatic amines, the nitroso group joins to the amino nitrogen atom to form an *N*-nitrosamine. Often potent animal carcinogens, the effect of these compounds on humans is being extensively studied. Their precursors are widely used in foods, such as cured meats, to prevent the growth of microorganisms responsible for the type of food poisoning called botulism. As nitrites are produced by reduction of nitrates in saliva and both nitrites and amines are formed by bacterial breakdown of proteins in the intestine, more nitrosamines are probably formed in the body than are eaten. As yet, no conclusive evidence of carcinogenic effects in humans exists for nitrites used in meat curing.

Four of the simplest types of organo-sulfur compounds are thiols (top, near right), which contain an —SH group; sulfides (bottom, near right), which have a sulfur atom bonded to two carbon atoms; sulfoxides (top, far right), with a sulfur atom doubly-bonded to an oxygen atom; and sulfones (bottom, far right), with a sulfur atom linked by two double bonds to two oxygen atoms.

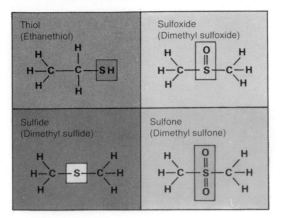

Thiol (Ethanethiol)

Sulfoxide (Dimethyl sulfoxide)

Sulfide (Dimethyl sulfide)

Sulfone (Dimethyl sulfone)

Organo-sulfur, phosphorus and metal compounds

Organic compounds containing sulfur are of increasing importance industrially, especially in pharmaceuticals and fundamental research. They also play a vital role in the metabolic processes that sustain life.

The simplest organo-sulfur compounds are thiols, or mercaptans, which are analogous to alcohols (their oxygen counterparts). They have the general formula RCH_2SH (an alcohol is RCH_2OH). But thiols are more acidic and more volatile than the corresponding alcohols and give off disagreeable odors. The characteristic smells of skunks, garlic and rotten cabbage are all attributable to various thiols.

The characteristic unpleasant smell of skunks (below) is mainly due to the organo-sulfur compound 1-butanethiol ($CH_3CH_2CH_2CH_2SH$), which is secreted by the animal. Thiols are also found in some plants; onions, for example, give off 1-propanethiol ($CH_3CH_2CH_2SH$).

Reactions of thiols

Oxygen and sulfur are both members of Group VIA in the Periodic Table, and so it is not surprising that many reactions are common to both alcohols and thiols. But there are slight differences. For instance, when hydroxyl (OH^-) groups are displaced from alcohols by more reactive substituents, they are often mopped up by surrounding protons (H^+ ions) in the reaction medium to form water. When thiol groups (SH) are displaced from thiols as SH^- ions, they are also mopped up by protons — to form hydrogen sulfide (H_2S) — but not nearly so readily as are hydroxyl ions. Neither do thiols react with reagents such as hydrogen bromide to give alkyl bromides. But thiols are easier to oxidize than are alcohols, giving a disulfide or, on further oxidation, a sulfonic acid. The formation of disulfides plays an essential role in living organisms. The amino acid cysteine contains a thiol group. Two of these cysteine thiol groups can be oxidized to form a disulfide "bridge" that links different parts of a protein chain, thus affecting the molecule's three-dimensional structure, which is crucial to its biological function.

Another useful and smooth reaction is the free-radical addition of thiols to alkenes (olefins) to form dialkyl sulfides. Thiols are easily prepared, either by adding sulfur to Grignard reagents, or by treating alkyl halides with sodium hydrogen sulfide (NaHS).

Other organo-sulfur compounds

Another valuable set of organo-sulfur compounds are the sulfoxides, which contain a sulfur-oxygen double bond; an example is dimethyl sulfoxide ($(CH_3)_2SO$), which is widely employed as a solvent for organic reactions.

Sulfoxides are usually produced by oxidizing organo-sulfides (which contain a sulfur atom bonded to two carbon atoms) using hydrogen peroxide. Further oxidation (also using hydrogen peroxide) yields sulfones, which contain two sulfur-oxygen double bonds. Some aromatic sulfones are used as monomers for plastics that are resistant to high temperatures.

Organo-phosphorus compounds

Comparison is inevitably made between organo-phosphorus and organo-nitrogen compounds, because both phosphorus and nitrogen are in Group VA of the Periodic Table and so are similar in character. But unlike nitrogen, phosphorus is unable to form multiple bonds with carbon, though it can form very strong double bonds with oxygen, and does so in many organo-phosphorus compounds, such as phosphine oxides ($R_3P=O$) and trialkyl phosphates ($(RO)_3P=O$). A valuable reaction involving an organo-phosphorus compound is the Wittig reaction, which can convert a carbonyl bond (C=O) (in, say, a ketone) into a carbon-carbon double bond (to give an alkene). The phosphorus agent responsible for the conversion is called an ylide (general formula $R_3P^+—CH_2^-$, where R is an alkyl or aryl group).

Further important properties of organo-phosphorus compounds include their ability to form adducts with acids, and their ability to form phosphonium ions (general formula R_4P^+) with organic halides.

Organo-phosphorus compounds also play an integral part in the transfer of energy within biological organisms. Adenosine triphosphate (ATP) is the most important carrier of energy in living cells. It releases energy — used to fuel the metabolic reactions that sustain life — by transferring one of its phosphate groups to another molecule, thereby becoming adenosine diphosphate (ADP).

Essential though phosphorus is in living systems, it can also threaten them. The nerve gases Sarin and Tabun are phosphorus compounds and they inhibit the enzyme acetylcholinesterase (which is essential in triggering nerve impulses). They do this by converting the hydroxyl group of a specific amino acid in the enzyme into a stable phosphate group. Some phosphorus-based insecticides act using the same mechanism. The principal ones are malathion and the parathions. A special feature of these insecticides is the speed with which they decay in the environment, so presenting no threat to people after use.

Organo-metal compounds

Certain types of compounds containing both alkyl groups and metals have been known for a long time: Grignard reagents (general formula RMgX, where R is an alkyl or aryl radical and X is a halogen) were first made around the turn of this century. They are often used to convert alkyl halides (RX) into hydrocarbons (RH):

$$RX + Mg \rightarrow RMgX \rightarrow RH$$

Organic compounds containing lithium are also well established, the best-known being butyl lithium. It can be used to "lithiate" other compounds to make reactive intermediates that are of great use in complex syntheses.

Other useful organo-metallics include aluminum trialkyls which, in conjunction with titanium salts, are important polymerization catalysts, and lead tetraethyl ($[C_2H_5]_4Pb$), an antiknock agent which until recently has been widely added to gasoline. It is now being phased out because of fears that the lead discharged in exhaust fumes may be harmful.

Scientists are now able to synthesize more exotic organo-metallic compounds called sandwich compounds, which are now finding uses as catalysts. They are so named because in many of them a transition metal atom, such as iron or chromium, is sandwiched between two planes of cyclic carbon compounds. The first to be synthesized (in the early 1950s) was ferrocene, an iron atom sandwiched between two molecules of cyclopentadiene; many more have since been made. Normally, the organic portion of such molecules is either an aromatic-type ring compound such as cyclopentadiene or benzene, or a compound containing multiple bonds, such as carbon monoxide (C=O) or ethene (CH_2=CH_2).

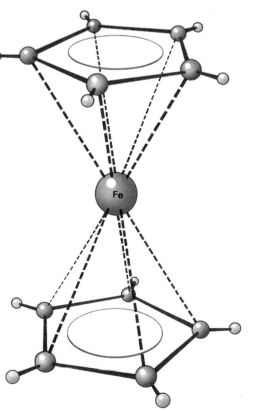

Sheep are dipped (above) in a liquid that contains organo-phosphorus compounds because these substances are particularly effective in destroying ticks and other external parasites.

Ferrocene is probably the best known of the group of organo-metal compounds called sandwich compounds. Structurally, it consists of an iron atom sandwiched between, and bonded to, two parallel cyclopentadiene rings — as illustrated in the diagram (left). Ferrocene can be prepared relatively easily — by treating cyclopentadiene with a Grignard reagent and then adding iron (II) chloride. At room temperature ferrocene is an orange solid and has a camphor-like smell. It is insoluble in water but dissolves in most organic solvents, such as ethanol and benzene.

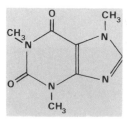

Caffeine (the structure of which is illustrated above) occurs in the beans of the coffee plant (right) and also in tea leaves, cola nuts and cacao. It is probably the best known of the alkaloids, a group of nitrogen-containing ring compounds that have physiological effects — caffeine, for example, is a stimulant and diuretic.

Monosodium glutamate (MSG) is commonly added to processed or cooked food — particularly Chinese food (below) — to enhance flavors. Normal salt (sodium chloride) must also be present for the monosodium glutamate to impart an attractive taste to the food.

Complex organic compounds

In addition to the comparatively simple compounds described in the preceding articles, organic chemistry includes thousands of complex compounds. Many of these occur naturally and are of importance in living organisms and as medicinal drugs. Others are synthetic — artificial organic chemicals for use in a wide range of products that includes drugs such as aspirin and pesticides such as DDT.

Alkaloids
More than 2,000 compounds have been classified as alkaloids. They all contain ring structures which bear at least one nitrogen atom, all show physiological activity, and most act as bases in chemical reactions.

All alkaloids are derived from plants; among the best known are nicotine from tobacco, opium and its derivatives (such as heroin and morphine) from poppies, caffeine from coffee, strychnine from the tree *Strychnos nux vomica,* and quinine from cinchona bark. Scientists have now discovered that all alkaloids are biosynthesized from amino acids such as tyrosine and tryptophan.

Synthetic pesticides
Pesticides are chemical agents used to kill animal or vegetable life-forms that harm agricultural produce. Those used to kill insect pests are in three main categories — chlorinated pesticides, organo-phosphates and carbamates. The most familiar chlorinated pesticide is DDT, 2,2-di(*p*-chlorophenyl)-1,1,1-trichloroethane, which like others in this group (including Aldrin and Dieldrin) is termed a "hard" insecticide because of its resistance to decay by the environment; it may persist for months or even years.

Organo-phosphates are the most powerful insecticides but are destroyed rapidly in the environment after use. The most lethal agents in this group are the parathions and malathion.

The carbamates are derivatives of carbamic acid (NH_2COOH). The most important of them is Sevin (1-naphthyl-*N*-methylcarbamate).

Because most insecticides are also lethal to man and other mammals, their use has aroused much concern among environmentalists. But scientists are now developing new ways to combat insect pests. One of the most effective is the use of pheromones. These are manufactured naturally by insects, which discharge them into the atmosphere to attract the opposite sex of the species for mating. Some insects can be attracted in this way from many miles or kilometers away. By making pheromones artificially, scientists can lure insects into traps. Or they can be used simply to confuse insects and disrupt mating, so severely reducing the insect population.

Other pesticides include herbicides and fungicides. The most widely-used herbicide is (2,4-dichlorophenoxyl)acetic acid (2,4-D) although (2,4,5-trichlorophenoxy)acetic acid (2,4,5-T) has attracted attention recently because it is claimed to pose a threat if used improperly. Fungicides are very diverse. They range from simple inorganic substances (such as lime and copper sulfate) to complex organic compounds of mercury and tin.

Synthetic and natural drugs
Drugs are chemical agents usually used to combat disease, to help to repair the body after disease or injury and to suppress pain, although some other uses — such as the building up of muscle tissue by athletes — do exist.

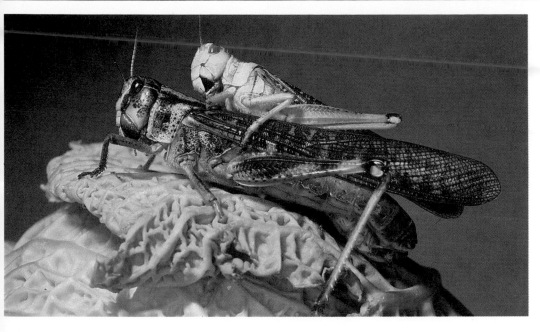

Locusts (left) and many other species of insects produce pheromones, complex organic compounds that the insects (usually females of the species) discharge into the air to attract mates. These substances are extremely potent — in many species, individuals respond after detecting only a few molecules. Chemists are now able to synthesize some pheromones, which offers the possibility of controlling insect populations without the extensive ecological disruption that can result from using conventional insecticides.

Some pure medicinal products can be isolated from natural sources, although such substances usually have to be modified chemically before they can be used. Other drugs are entirely synthetic, an example being aspirin, 2-(acetyloxy)-benzoic acid, which is now produced from phenol. Perhaps the best-known of all drugs is the antibiotic penicillin — first discovered in 1928 by Alexander Fleming — of which several different forms exist. Nowadays, it is produced industrially by isolation from mutant strains of the mold *Penicillium chrysogenum*. Penicillin combats various cocci and spirochaetes although it is less effective against rod-shaped bacteria. Different penicillins can be synthesized by altering the chemical composition of the growth medium used to sustain the mold.

Other important antibacterial drugs are the "sulfa" compounds, which all contain the sulfanilamide grouping in their structures. They act by competing with *p*-aminobenzoic acid, a substance that some microbes require in order to thrive. Removal of the nutrient by the sulfa drug starves and kills the bacteria, and cures the patient.

A newly-developing way to produce drugs is by manufacturing special bacteria or plant cells which act as tiny "drug factories." They can be modified genetically to produce important drugs. Insulin and interferon are already being produced commercially by this novel technique, and more products are sure to follow as research into biotechnology intensifies.

Artificial sweeteners and flavorings
By far the best-known artificial sweetener is saccharin, although some experiments with animals have led to the suspicion that it may cause cancer. Other synthetic sweeteners, sodium and calcium cyclamate, were banned in 1970 when they were found to cause bladder cancer in rats. Scientists are now developing new sweeteners. One, 6-chloro-D-tryptophan, is nearly a thousand times sweeter than ordinary sugar (sucrose). Others include Acesulfame-K, which does not accumulate in the body, and aspartame, a dipeptide. About 800 synthetic food flavorings are now also in use. A common flavor enhancer is monosodium glutamate, which is usually isolated from natural sources such as flour or soybean, or produced by fermentation and then purified.

Chelating agents
These are organic groupings that are capable of linking to metals to form coordination compounds which, in some cases, are quite bulky. They can be useful in removing metallic poisons, such as lead, from the body.

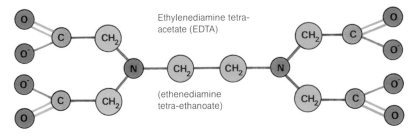

Ethylenediamine tetra-acetate (EDTA)

(ethenediamine tetra-ethanoate)

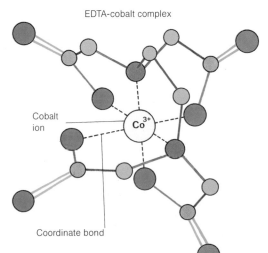

EDTA-cobalt complex

Cobalt ion

Coordinate bond

Chelating agents, such as EDTA (ethylenediamine tetra-acetate) illustrated above, can form coordinate bonds (a type of covalent bond) with metals, thereby forming coordination compounds, also called complexes. The diagram left shows how a molecule of EDTA wraps around and forms coordinate bonds (the dotted lines) with a cobalt ion (Co^{3+}).

Color in organic chemistry

Many organic compounds are brightly colored. Some, such as the green plant pigment chlorophyll, occur naturally whereas others are synthesized by chemists for use as dyes and colorants. A feature shared by all colored organic compounds is the existence in their structures of sequences of atoms linked by double bonds — either in chains or in rings. It is these sequences —

or chromophores — that are responsible for absorbing specific wavelengths of light. Wavelengths that are reflected make the compounds appear colored, and so by altering the number and sequence of doubly-bonded atoms, chemists can deliberately vary the colors of compounds.

Natural pigments

Many of the naturally-occurring organic pigments show similarities in their molecular structures: they are all aromatic compounds, and have oxygen atoms in their structures. They account for most of the yellow, red and blue colors in flowers and fruits, and may be divided into various subgroups. The catechins, based on catechol, are used as dyestuffs and in special inks. Although colorless, they turn brown on exposure to air and light.

Another group are the leucoanthocyanidins, of which indigo — the dye used in blue jeans — is probably the best known. It can be prepared easily from a compound called indican which exists in the indigo plant *(Indigofera tinctoria),* and this accounts for the wide use of indigo as a dyestuff since ancient times.

A third group contains flavonols, flavones and anthocyanins. Flavonols and flavones account for the ivory and yellow colors of many plants, whereas anthocyanins are responsible for most of the red and blue colors. The particular color imparted by an anthocyanin compound depends on the acidity or alkalinity in the plant; for example, cyanin, the pigment in red roses and blue cornflowers, is red in dilute acid and blue in dilute alkali.

Another group includes the naphthoquinones and anthraquinones. Probably the most familiar example of the former is lawsone, the substance in the leaves of the henna shrub *(Lawsonia* sp.) that gives the orange-red color to henna dye. Carminic acid is a typical anthraquinone; it is the principal pigment in cochineal, the scarlet dye obtained from the dried, pulverized bodies of the insect *Coccus cacti* widely used in food and cosmetics.

Other common natural pigments include chlorophyll, carotene and rhodopsin. Chlorophyll plays a key role in photosynthesis, the process by which plants harness the energy of the Sun to manufacture nutrients. Carotenoids are responsible for the orange, yellow and red pigments in many flowers and animals. A well-known example is beta-carotene, which imparts the orange color to carrots. Rhodopsin, otherwise known as visual purple, occurs in the retina of the eye and isomerizes when subjected to light, thus triggering the visual process.

Azo dyestuffs

Azo dyes are distinguished by the presence in their structures of at least one nitrogen-nitrogen double bond, usually linked to various aromatic

Indigo and Tyrian purple have both been used as dyes since ancient times. Indigo was originally obtained from various species of the indigo plant *(Indigofera* sp.), principally from *Indigofera tinctoria* (top right), which is indigenous to parts of eastern Asia and Central America. In the late 1870s, however, the dye was synthesized and today almost all indigo is obtained in this way. The structure of indigo is very similar to that of Tyrian purple; the only difference is that Tyrian purple contains two bromine (Br) atoms — as can be seen by comparing the structures in the diagram (middle right). The presence of these bromines has the effect of displacing the color of the compound toward the red, resulting in the purple color of Tyrian purple as opposed to the dark blue of indigo. Like indigo, Tyrian purple was originally obtained from a natural source, the aquatic mollusk *Murex brandaris* (the shell of which is shown below), but has today been superseded by synthetic dyes.

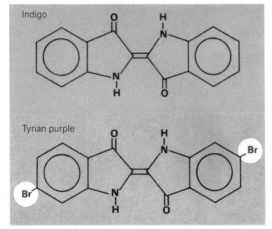

Indigo

Tyrian purple

groups. They are normally synthesized by adding diazonium salts (which are formed by the action of nitrous acid on aromatic amines) to aromatic compounds such as phenol or aniline. Perhaps the best-known azo dye is Congo Red, a brown-red compound. Another is Butter Yellow, a vibrant orange substance which was used as a food additive until suspected of causing cancer.

Many of the most widely used azo dyes contain sulfonate groups in their structures. These make the dyes soluble in water and further serve to bind them tightly to the large polymeric molecules of textiles. Because some azo dyes change color when reacted with an acid or base, they are used by analytical chemists as indicators in acid-base titrations. The color change is abrupt and is caused by alteration of the double bond sequence in the molecule when it is converted from an acid into a base or vice versa.

Metal-organic dyes

A number of important metal-organic dyes are based on the compound phthalocyanine, a blue-green compound which itself is not bound to a metal. Binding phthalocyanine to copper gives a striking blue dye, which is used in paints, printing inks, resins, colored chalks and pencils. By replacing hydrogen atoms in copper phthalocyanine with chlorine atoms, it is possible to form a bright green compound; called chlorinated copper phthalocyanine, it serves much the same uses as copper phthalocyanine Another green compound can be synthesized by replacing two hydrogen atoms in copper phthalocyanine with sulfonic acid groups ($-HSO_3$). Other uses for these dyes include decorative enamels and paints for motor vehicles.

Food colorants

To be accepted for use as a food colorant, a compound must satisfy strict regulations relating to its toxicity. Indeed, many compounds formerly used as food colorants have now been banned because they fail to satisfy current regulations. An example is amaranth ("Red No. 2"), the trisodium

salt of 1-(4-sulfo-1-naphthylazo)2-naphthol-3,6-disulfonic acid, which at one time was the most widely used food colorant. It is now thought to be a carcinogen, and as such has been banned from use in food by the United States Food and Drug Administration.

The brilliant orange-yellow robes of Buddhist monks (below) are traditionally dyed using the dried stigmas of the saffron plant, a species of crocus *(Crocus sativus);* the active consti-tuent of this plant is a volatile oil called picrocrocin. In addition to being employed as a coloring agent, saffron is also used for flavoring, imparting a bitter, aromatic taste to food.

Inks are available in a wide range of colors — as exemplified by the felt-tip pens above — and many of the dyes used in them are synthetic organic compounds, principally azo dyes and metal-organic dyes. The former are derivatives of azobenzene ($C_6H_5N=NC_6H_5$) and are usually red, yellow or brown. The main metal-organic dyes are synthesized from the blue-green compound phthalocyanine.

Organic synthesis

Organic chemists can construct complex organic compounds from simpler ones using a variety of reactions — the techniques of organic synthesis. To make just one compound, a chemist may have to perform a long series of reactions using many different reagents. There may also be several ways of making any given compound.

In choosing which synthetic route or pathway to adopt, the chemist has to compromise between which reaction procedures are simplest and most convenient, which reagents are cheapest or most widely available, and which reactions give the best yields of compounds to be used in the next step of the synthesis.

The reasons for doing a synthesis are threefold. One is that particular organic products may be in demand commercially. Examples are drugs, which are often synthesized by carefully constructed and elegant synthetic routes. Another reason is that a chemist may wish to establish the structure and identity of a natural product. A great many of these are isolated from plants and animals, and because they may possess valuable properties — perhaps medicinal or industrial — it is important for scientists to know their structures. They do this by synthesizing compounds — often with much trial and error — that correspond exactly with the natural product, both in chemical reactions and physical properties. If the compound is shown unambiguously to be identical with the natural product, then this confirms what its structure is. The third reason is to make completely novel compounds, either to test theories of reactivity in chemistry, or because the new compound may turn out to have important properties.

Strategies of synthesis

Once a chemist has decided upon the need to synthesize a product, he has to work out a satisfactory synthetic route by which to make it. The usual strategy is to work backwards from the required product. For instance, if a complex product Z is required, it may be possible to make it by modifying a precursor, Y. This in turn may be available from X, and so on, until the chemist reaches starting materials that are readily available and cheap. But because each compound on the way to the final product can usually be synthesized by a variety of methods, the scope for developing an array of reaction pathways is almost limitless. Essentially, the chemist usually has to change one part of the molecule at a time, while leaving the rest of it undisturbed. So in deciding upon which synthetic route to choose a number of factors have to be weighed against one another. The route that uses the cheapest and most widely available starting materials might seem to be preferable, especially in industrial synthesis. But this is not always the most convenient route from a chemist's point of view because it may involve dangerous reactions, or perhaps some of the steps give only small amounts of the product needed for the next stage of the synthesis (the so-called intermediates). This also means wastage of starting materials, although in some industrial syntheses it may be possible to recycle starting materials and intermediates so that the effective yield is increased.

The table lamp (below left) dates from the late 1920s and is made of Bakelite, the first successful thermosetting plastic resin to be produced (in the early 1900s). It is formed by the condensation reaction between phenol (usually synthesized from benzene) and methanal (formaldehyde); the resultant phenolic resin is then melted and allowed to set, a process that makes the material extremely hard and unmeltable on further heating. Phenolic resins such as Bakelite are only one of the many types of chemicals that can be synthesized from benzene; some of its other important synthetic derivatives are shown in the diagram (below right).

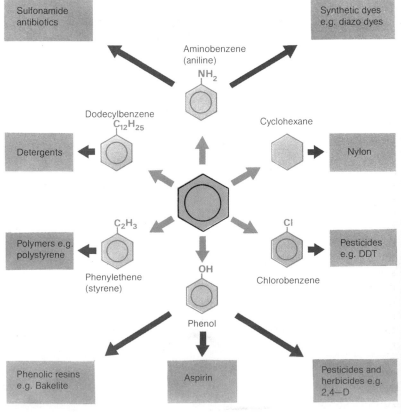

The amount produced in each stage is indicated by the yield, the percentage of starting material in that particular step which is converted into the required intermediate. Other problems might arise in purification, because starting materials and unwanted coproducts may contaminate the product. So if the intermediate yields are low, the synthesis as a whole may be expensive.

A further drawback may be that some of the steps proceed very slowly. The chemist may try to speed them up by raising the temperature of the reaction or by using a catalyst, but care is needed to avoid destroying the product, or synthesizing other, unwanted substances.

Synthetic procedures

Once the chemist has decided which route to follow, he has to anticipate where problems might arise. Each step must be examined to make sure it will proceed as planned. Also, the chemist must consider whether a particular step might affect the wrong part of the molecule. If this is likely, he may be able to shield the molecule with what is termed a "protecting group" which temporarily denies reagents access to groups in the molecule that are prone to attack. It can be removed to restore the original grouping at a later stage.

The chemist also has to consider which steps may be impossible for practical reasons. For example he may be unable to achieve the pressures needed for one stage, or the reaction may be too explosive to risk in the laboratory. The more complex the product required, the more problems face the organic chemist because as the number of steps in a reaction increases, so too do the number of options open to him. Also the problems of unwanted side reactions multiply as the size of the molecule — and hence the number of sites prone to attack — increases. It is therefore not surprising that many satisfactory synthetic routes to complex products take years to perfect.

New developments

Advances in computer technology are making synthesis much easier. Details of synthetic methods, of alternative methods of interconverting chemical groupings, and of the suitability of reactants for given conversions can all be stored and processed by computers. The chemist can then choose the optimum route available on the basis of computer predictions.

Computers are also valuable in scaling-up and processing operations. Traditionally, scale-up follows a familiar pattern. Having established the need to make a particular product, one or two synthetic routes are investigated. The most favorable route is then developed into a process that can be carried out in a production environment.

Besides assisting the chemist in formulation and choice of route, computers can be used to regulate and monitor reaction conditions. Remote operation provided by a computer can also permit safe handling of toxic raw materials, intermediates and products, and the utilization of hazardous reactions.

Protecting groups play an essential part in many organic synthesis reactions, as demonstrated in the example (above). It is impossible to convert compound 1 into compound 4 directly by simply adding a hydroxyl group (OH) at A because the hydroxyls would add preferentially to the double bond at positions B and C. These vulnerable sites are therefore protected by adding bromine (Br), to give compound 2. The hydroxyl group can then be added at the desired site, forming compound 3. Finally, the bromine atoms are removed to restore the double bond and give the required compound (4).

Polymers

Polymeric materials are composed of very large molecules (macromolecules) formed by linking together many smaller, more simple units called monomers. There can be as few as five or as many as several thousand monomer units in a polymer. Typical examples of polymers are plastics: polyethylene film, a transparent material used in packaging; polyurethane foam made into cushions and mattresses; and nylon and polyester fibers used in textiles. Synthetic resins for paints and adhesives are also polymers. Rubber is a natural polymer isolated from trees native to South America (but now grown mostly in Asia), although more than half the rubber used today is synthetic. Other natural polymers include various protein substances and plant carbohydrates such as starch and cellulose.

Scientists have a broad range of monomers at their disposal and can modify them, and the polymerization conditions, to make synthetic polymers which suit particular needs and exhibit special properties. These can vary also depending on the number of monomer units in each macromolecule and the way they link together. Branched-chain polymers are normally stronger than straight-chain polymers, because of the extra linkages between chains. They can also respond differently to the action of heat — thermoplastic polymers can be molded again if remelted, whereas thermosetting polymers harden permanently once molded.

If a single substance is used as monomer, the resulting polymer is called a homopolymer, whereas copolymers are formed by linking together different kinds of monomers. Reactions to splice monomeric units can also vary. In con-densation polymerization, a small molecule such as water is ejected by monomers as they come together, whereas addition polymers are formed by direct linkage of monomer units.

Polyethylene

Perhaps the most familiar of all polymers, polyethylene exists in various forms according to the way it is manufactured. Low-density polyethylene (LDPE) was the first type to be produced, by heating ethene to about 200—600°F (about 100—300°C) under a pressure of 1,000—2,000 atmospheres. During polymerization, some chains branch slightly so the molecules cannot pack closely — hence its low density. It is a soft substance and can be used as packaging material and for plastic bags, squeezy bottles and so forth. By contrast, high-density polyethylene (HDPE) is composed of close-fitting chains, which make it ideal for blow-molded and injection-molded products. It is made by subjecting ethene in solution to a pressure of 5—30 atmospheres in the presence of a Ziegler catalyst (containing, for example, aluminum and titanium). Low-molecular weight polyethylene (LMWPE) has good electrical resistance, plus excellent resistance to chemicals and abrasion. It is used in paper and container coatings, liquid polishes and textile finishing agents. Its molecular weight ranges between 1,000 and 10,000, whereas for the other varieties of polyethylene it is generally between 10,000 and 400,000 and may reach as much as 1,000,000.

Polypropylene

Polypropylene is a thermoplastic polymer whose molecular weight is usually higher than that of polyethylene. There are three forms, which differ according to the arrangement in which the monomer units are put together. The order has a

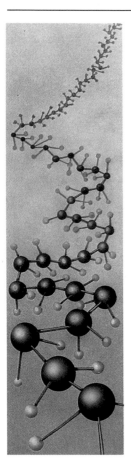

Polymers are giant molecules formed from many small units. One of the simplest is polyethylene, which consists of long chains of carbon atoms (black), each carrying two hydrogen atoms (yellow).

One of the main methods of polymerization involves a chain reaction triggered by the presence of a free radical — a highly reactive species bearing an unpaired electron (shown by the blue dot). The usual source is an alkyl peroxide, which decomposes into free radicals on heating. These combine with the monomer to form more free radicals, which in turn react with more monomer molecules to build up a long chain. Finally two long-chain components combine to form a molecule of polymer and so terminate the reaction.

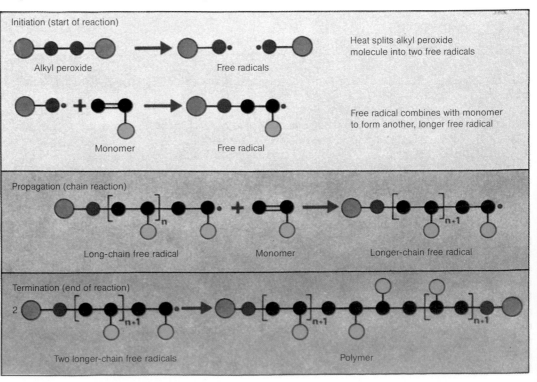

Initiation (start of reaction)

Alkyl peroxide   Free radicals

Heat splits alkyl peroxide molecule into two free radicals

Monomer   Free radical

Free radical combines with monomer to form another, longer free radical

Propagation (chain reaction)

Long-chain free radical   Monomer   Longer-chain free radical

Termination (end of reaction)

Two longer-chain free radicals   Polymer

marked effect on the properties of the plastic and accounts for its wide range of uses, from carpets to containers. By employing special Ziegler catalysts, polymer scientists are able to link the monomeric propene (propylene, $CH_3CH=CH_2$) units in regular sequences, which impart more mechanical strength to the polymer. The highly stereospecific isotactic form has a crystalline structure and is used in a variety of blow-molded items and for packaging film, whereas atactic (disordered) polypropylene, being more pliable and semitacky, can be used with bitumen on roofs and roads and in backings for carpet tiles.

Polystyrene

This material, renowned for its use as a thermal and electrical insulation, is a thermoplastic resin made by polymerizing an emulsion of phenylethene (styrene) using a peroxide catalyst. Perhaps most familiar as rigid foam or as expandable beads, it is used in packaging, ice buckets, water coolers, furniture construction, thermal insulation, ceiling tiles, and a variety of other items. The most common use for solid polystyrene is in making ball-point pens. A special property of branched — as opposed to linear — polystyrene is its insolubility. Instead of dissolving in a solvent, it swells up, the degree of swelling depending on the degree of cross-linking. It thus forms a resin.

Halogenated polymers

Polyvinyl chloride (PVC) became well-known in the 1960s when it was first used as a material for plastic raincoats and simulated leather goods. It is manufactured in a similar way to polystyrene, again with a peroxide catalyst. To make it soft enough for clothing and simulated leather, it has to be mixed with a plasticizer, or softener, such as dioctyl phthalate, an aromatic ester. In its harder form, it is used in pipes, panels, and other molded parts. Its use has declined slightly since the monomer, chloroethene (vinyl chloride, $CH_2=CHCl$), was shown to cause a rare form of liver cancer if inhaled. A well-known fluorine-containing polymer is Teflon (polytetrafluoroethylene or PTFE). It is a very low-friction, slippery material which makes it ideal for coating cooking pans to stop food sticking to their surfaces. Its manufacture involves polymerizing tetrafluoroethene (tetrafluoroethylene, $CF_2=CF_2$) in water, under pressure, using ammonium persulphate as a catalyst.

Natural and synthetic rubbers

Rubbers are special polymers with elastic properties — that is, they return to their original shape after being deformed or stretched. Natural rubber, obtained from the latex of *Hevea* trees, is a polymer of isoprene (2-methylbuta-1,3-diene, $CH_2=C(CH_3)CH=CH_2$). It is used in cements, adhesives, tape for insulating electrical equipment, and for cable wrapping. It becomes more useful when vulcanized, a process in which sulfur reacts with the carbon-carbon double bonds to form "bridges" of sulfur atoms which cross-

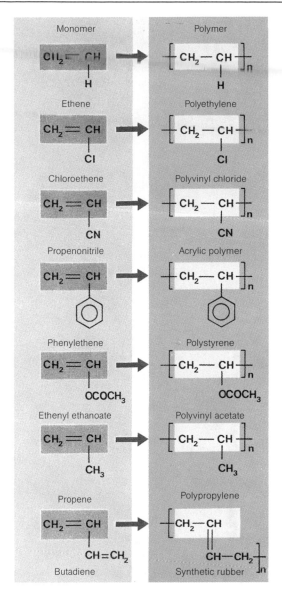

Polyethylene — made by polymerizing ethene or ethylene ($CH_2=CH_2$) — was one of the first polymers, and many others can be best understood by considering their monomers as derivatives of ethene. In this diagram, the chemical formulae of the monomers (left hand column) have been presented so as to emphasize this common factor, with the double-bonded "ethene" parts of the molecules on pink panels. After polymerization, this element of the molecule (but now only singly bonded) forms the long repeating backbone or chain of the polymer, shown on yellow panels in the right-hand column.

Rubber is valued for its resilience and elasticity, two essential properties for the inflatable children's "castle" (below). Once only a natural product, most types of rubber are now synthetic polymers.

Thousands of logs are stored to meet the voracious appetite of a pulp mill. Wood contains cellulose (see diagram, opposite), a versatile organic substance which will probably gain even greater importance with the decline in reserves of petroleum and the petrochemicals from which most polymers are currently manufactured.

link the molecules. Vulcanized rubber is made into tires (principally), hoses, footwear and a variety of other items. Important synthetic rubbers are SBR (styrene-butadiene rubber) and nitrile rubber (acrylonitrile-butadiene), both of which are copolymers, and chloroprene rubber, which is made from the isoprene analogue 2-chlorobuta-1,3-diene (chloroprene).

Polyamides

This group of high molecular weight polymers is characterized by the amide (—CO—NH—) linkages which bind the monomers together. Formed by condensation reactions in which an organic acid and an amine add together, they may be natural (as in proteins) or synthetic. By far the best known synthetic polyamide is nylon, which was first discovered shortly before World War II when scientists at Du Pont in the United States were searching for silklike polymers. The original form was made from hexanedioic (adipic) acid and 1,6-diaminohexane (hexamethylenediamine). After all the water has been driven off by heat, the polymer is extruded in a ribbon from the polymer melt, or spun into fibers which may be woven to make textiles. Different types of nylon can be produced by altering the numbers of carbon atoms in the monomers. One modern form uses only a single monomer — with an amino group at one end and a carboxylic group at the other.

Other synthetic polyamides are the aramids — aromatic polyamides — renowned for their flame-retardant properties. One of them, Kevlar, was specially developed for use in the belting for radial motor tires.

Polyesters

Also condensation polymers, polyesters are formed by reacting organic acids with polyfunctional alcohols. Polyethylene terephthalate (PET) is the best-known of all polyesters, being used as a textile fiber under various trade names such as Terylene and Dacron. PET can also be formed as film (Mylar) and, in recent years, has become important as the raw material for plastic containers for soft drinks.

Polyurethanes

Again condensation polymers, polyurethanes are manufactured by adding diisocyanates to polyfunctional alcohols, such as butane-1,4-diol.

The best-known of all polyurethanes is polyurethane foam, which is prepared by adding water to the reaction mixture. It reacts with the diisocyanate to form cross-links in the polymer and to liberate carbon dioxide gas, which causes foaming. Water is thus said to act as a blowing agent. Flexible foams are made using polyoxypropylene diols or triols, whereas rigid foams are manufactured using polyethers. The foam has become so familiar since it has been widely used in furniture, especially as mattresses and as seat cushions.

Polyurethanes are also excellent thermal insulators, used to shield storage tanks and ships' holds and to lag pipes.

Cellulose acetate and nitrate

The base material for these polymers is the natural polymer cellulose, which consists of long molecular chains formed by condensation of glucose molecules. It is the structural material of plants, the major component of wood pulp, paper and cotton, and one of the cheapest and most abundant organic substances available. Wood pulp contains about 1,000 subunits per macromolecule, and cotton fibers can contain up to 3,500. Cellulose acetate is made by adding acetic anhydride or ethanoic (acetic) acid to cellulose derived either from wood pulp or cotton. It can be spun into yarn to give the widely-used textile fabric rayon acetate. It is also employed as a film base, and as a plastic.

Cellulose nitrate, or nitrocellulose, was one of the earliest polymers known. It is made by treating cellulose with a mixture of nitric and sulfuric acids under carefully controlled conditions. Alcohol groups in the cellulose are converted into nitrate ester groups. By varying the reactant proportions, it is possible to regulate the number of these formed. Guncotton, an explosive, is highly nitrated. The low-nitrated form, pyroxylin, is used in fast-drying lacquers, and in table-tennis balls. Its use for cine film has been discontinued because of its high inflammability.

Resins

Natural resins, such as water-insoluble mixtures of high molecular weight carboxylic acids, are soft and pliable when warm, and hard and brittle when cool. Synthetic polymers, such as polysty-

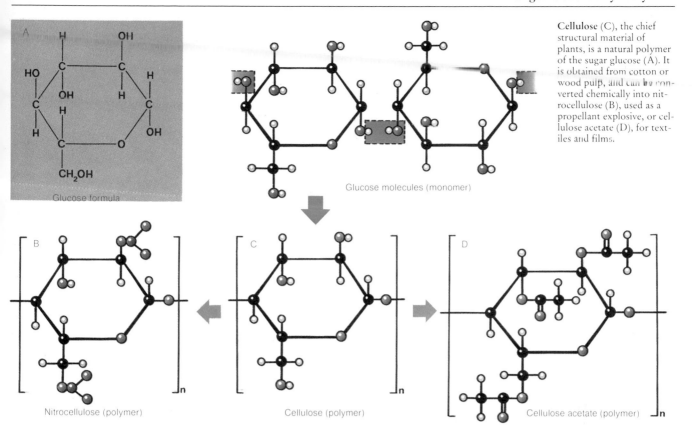

Glucose formula

Glucose molecules (monomer)

Cellulose (C), the chief structural material of plants, is a natural polymer of the sugar glucose (A). It is obtained from cotton or wood pulp, and can be converted chemically into nitrocellulose (B), used as a propellant explosive, or cellulose acetate (D), for textiles and films.

Nitrocellulose (polymer)

Cellulose (polymer)

Cellulose acetate (polymer)

rene, rarely display these types of characteristics and so to call them resins is somewhat misleading. They often exhibit good chemical and electrical resistance, and are normally insoluble in water. In addition, they are easy to process by molding, machining or extrusion. Epoxy resins, formed by combining epoxides such as epichlorohydrin with a hydroxyl compound such as bisphenol A, have a broad range of uses, the major one being as a strong adhesive. They also form tough protective coatings and are employed in corrosion-resistant paints for use on metals.

Silicones

Useful polymers which exist as oils, resins and rubbers, silicones are formed by adding chlorosilanes to water. Their value lies in their inertness. Untouched by water, chemicals, high and low temperatures, and electricity, they are usually based on chains of dimethyl silicon groups linked by oxygen atoms. The rubbers can be used as seals and gaskets at extreme temperatures, the resins as electrical insulators and water-repellent paints, and the oils as lubricants in extreme conditions of temperature.

Low-density polyethylene film can be made by blowing hot air through a polyethylene tube. The heat softens the plastic and the air pressure inflates the tube, stretching it like a balloon to form a thin film which is used mainly for making plastic bags and wrapping material.

Fact entries

Polyalkenes are the major group of synthetic polymers. They include polyethylene and polypropylene (made by polymerizing ethane and propene), and the closely related (chemically) polyvinyl chloride (PVC), polytetrafluoroethylene (PTFE) and polystyrene, which are polymers of, respectively, chloroethene (vinyl chloride), tetrafluoroethene and phenylethene (styrene). Most synthetic rubbers are polymers based on butadiene and its analogues.

Polyamides, best-known in the various types of nylon, are condensation polymers; for example Nylon 6/6 (each of its monomers contains six carbon atoms) is made by reacting hexanedioic acid and diaminohexane.

Polyesters include polyethenyl acetate (polyvinyl acetate, PVA), used in emulsion paints and adhesives, and polymethyl but-2-enoate (polymethyl methacrylate), a plastic that can be made optically clear for lenses, etc. Polyethylene terephthalate (PET) is used for fibers and films.

Biochemistry

Biochemistry is concerned with the chemistry of living things. During the twentieth century and particularly in the postwar period, scientific knowledge about the working of living systems at the molecular level has increased dramatically, and this field is now generally treated separately from the rest of chemistry. A part of the subject, concerned with the way in which genetic information is coded and used by organisms, has had such an explosive growth in recent decades that it is often regarded as a separate science: molecular biology.

Biochemistry concerns itself with more than just the structures of molecules found in living systems. It also studies how such molecules are produced, what changes they may undergo in living cells, how they interact with different parts of an organism, what chemical events underlie the effects which they consequently bring about, and what happens to them subsequently. Thus, the process of digestion is a proper study for biochemistry: what sorts of molecules are taken in by an organism in its food, how does it break them down, and what does it do with the end products? Biochemistry is also concerned, in this example, with an important intermediate product — energy — which the organism obtains from food in order to sustain itself. How it extracts this energy from the food substances and how this is utilized in the energy-requiring processes of life are among the questions that biochemists have tackled, as is the fundamental question of how the energy of sunlight is trapped by green plants in the creation of food substances by the process of photosynthesis.

Large and small molecules

Looked at from a traditional chemical viewpoint, biochemistry can be divided roughly into two parts, one concerned with large molecules, the other with small ones. In organic chemistry, most of the familiar polymers are made up from hundreds of repeating units of one or two different small molecules. In living systems, three main types of polymers occur: polysaccharides, proteins and nucleic acids. The polysaccharides tend to be like the industrially important polymers in that they are generally made up of only one or two different monomers. Proteins, however, are made up from a much greater number of monomers. Although the basic core of the monomer is the same, there are differences in side chains which lead to the immense variety of proteins found in living organisms.

The structures of proteins are stored in coded form by the nucleic acid polymers ribonucleic acid (RNA) and deoxyribonucleic acid (DNA). Although the nucleic acids also have a simple, repeating backbone, they use only a few different side chains, which are arranged in an extremely precise way. By contrast, most commercial polymers which incorporate more than one monomer have only an average structure. Thus, although two molecules of monomer A may be incorporated for each three molecules of monomer B in a batch of a polymer, a short length of polymer chain still might have the structure —AABABBAAAB—, in which this ratio is reversed. Only the overall composition of the batch, which contains a large number of individual molecules, has the correct ratio of A and B.

The control of living systems needs to be much more exact than a statistical average. Consequently, the natural synthesis of nucleic acid polymers is controlled very carefully. The DNA strand that makes up a given gene in an organism, for example, always has the same structure, because of the individual way in which it is made, by copying an existing molecule a monomer at a

Digestion breaks down the food we eat into forms in which it can be absorbed through the walls of the alimentary tract. The main sites of digestion in humans are the mouth, stomach and small intestine; the table (below) gives the principal digestive reactions that occur in each.

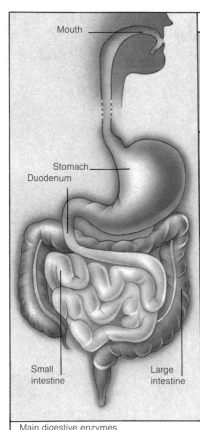

	Digestive actions (hydrolysis)
Mouth	**Mouth** Polysaccharides to disaccharides e.g. $(C_6H_{10}O_5)_n \rightarrow C_{12}H_{22}O_{11}$
Stomach Duodenum	**Stomach** Proteins to polypeptides e.g. $H(NHRCHCO)_mOH \rightarrow H(NHRCHCO)_nOH$
Small intestine	**Small intestine** Fats to fatty acids and glycerol e.g. $[R(CH_2)_nCOOCH]_3H_2 \rightarrow R(CH_2)_nCOOH + (CH_2OH)_2CHOH$ Polypeptides to amino acids e.g. $H(NHRCHCO)_nOH \rightarrow NH_2RCHCOOH$
Large intestine	Disaccharides (e.g. maltose) to monosaccharides (e.g. glucose) e.g. $C_{12}H_{22}O_{11} \rightarrow C_6H_{12}O_6$

Main digestive enzymes	
Mouth Amylase (polysaccharides)	
Stomach Pepsin (proteins)	
Small intestine Lipase (fats) Peptidase (polypeptides) Maltase (disaccharide) Amylase, lactase and sucrase (polysaccharides) Trypsin (proteins)	**Also** Polysaccharides to disaccharides Proteins to polypeptides

time. Where the gene ceases to be a precise copy, however, the result may be fatal for the organism concerned. The role of genetic mutation, as such change is called, in the development of cancer cells is an example.

Control mechanisms

The maintenance of life frequently depends on complex series of controls. Many of the small molecules important in biochemistry carry out control functions. Examples of such biochemical controllers are hormones.

Small molecules are also needed for other functions. The ability of certain metals to catalyze (bring about or speed up) chemical reactions is apparent from a study of both inorganic and organic chemistry. Life would be impossible without catalysis and, during evolution, the particular catalytic properties of some metals have been exploited. For a plant to capture the energy of sunlight, magnesium is essential; and the oxygen we need to survive is captured for us with the help of iron atoms. In each case, however, the metal ion is held in a complex molecule, which is itself associated with an even more complex biochemical system.

More than 600 years ago, it was realized that certain diseases could be cured by changes in the diet. But it is only during the twentieth century that the relatively small molecules responsible — the vitamins — have been analyzed chemically and have been shown, in many cases, to be low molecular weight substances which can combine with specific proteins to form powerful catalytic agents called enzymes. Many enzymes, however, are composed of protein alone, with no vitamin-derived "coenzyme."

Interactions

Catalysts — whether biological or not — work by lowering the activation energy of a reaction. They provide a more favorable environment for a reaction to take place. Complex biological catalysts — enzymes — also require a favorable environment to work effectively. Much of biochemistry is concerned with the environment in which molecules in living systems are found and with the ways this affects their behavior.

Just as chemistry provides a more detailed account of the composition of materials than does classical physics (which is concerned mainly with their overall physical properties), so biochemistry involves an additional factor. Inorganic and, despite its name, organic chemistry deal with inanimate materials; biochemistry studies reactions in living organisms.

Thus, a study of proteins and lipids (fats) in isolation can provide considerable information about them. But it requires a biochemical study of their interactions to help to explain the properties of membranes in living cells, which are made up from an intimate and organized physical combination of the two types of molecules.

Biochemistry is the chemistry of life, some of the most significant features of which are illustrated in the photograph above. The starling *(Sturnus vulgaris)* is feeding, and there is also evidence of the products of digestion. Flight requires energy, which comes from the biochemical processes of metabolism. Green plants are the means by which solar energy is converted into an edible energy source. And the pigments that produce plant and animal colors are also created biochemically.

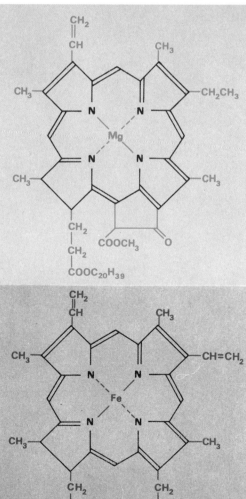

Two porphyrins, the plant pigment chlorophyll (top left) and the blood pigment heme (bottom left), play vital roles in the biochemical processes of photosynthesis and respiration respectively. Porphyrins are organic pigments that form complexes with metal radicals, which determine the specific properties of otherwise similar complex molecules.

Polysaccharides and sugars

Carbohydrates such as starch and sugar are a vital source of fuel for both plants and animals. They are found in much greater concentrations in plants, because they are also used in building cell walls. Bacteria also have carbohydrate-based cell walls and although animals are less dependent on the structural properties of carbohydrates, they perform a number of vital functions either alone or combined with proteins and other molecules.

The name carbohydrate refers to the general formula $(CH_2O)_n$, which suggests that they are hydrates of carbon. Most carbohydrates conform to this general formula (for example glucose is $C_6H_{12}O_6$) but some do not, and some also contain other elements such as nitrogen or sulfur. The simplicity of this formula also disguises the fact that there is a tremendous variety of carbohydrates, with subtle differences in structure which affect their properties and distribution in nature.

The simplest of carbohydrates are the monosaccharides (the name comes from the word for sugar in Greek). They are single units made up of unbranched carbon chains between three and seven atoms long. The most common are the trioses (three carbons), pentoses (five) and, the best known ones, the hexoses (six). These single units can join together to form the other major carbohydrate groups, the oligosaccharides, which can contain as few as two monosaccharide units (for example, sucrose), and the polysaccharides (for example starch), which may contain many thousands.

Structure
All simple carbohydrates contain a chain of carbon atoms, each joined to a hydroxyl (—OH) group, and one carbon atom linked to an oxygen atom by a double covalent bond. This may be at the end of the molecule as in an aldehyde, in which case it is called an aldose. Alternatively, it may be similar to a ketone, with the double bond in the middle of the chain; it is then known as a ketose. Glucose and fructose are both hexose sugars, but the former is an aldose and the latter a ketose.

The number of atoms alone does not determine the structure of the molecule. Apart from dihydroxyacetone, a simple triose sugar, all monosaccharides contain at least one asymmetrical carbon atom. Because of its tetrahedral shape a carbon attached to four different groups can be arranged into two separate formations which are mirror images of each other, a phenomenon known as stereoisomerism. Because sugars may have several asymmetrical carbon atoms, nomenclature can be confusing, so sugars that are complete mirror images are given the same name, but with a prefix of D or L, depending on their orientation.

Nearly all monosaccharides in nature are found in the D form. Most monosaccharides are also optically active, in that the individual asymmetric carbon atoms rotate the plane of polarized light to the right or left. The symbol (+), or sometimes a small d, indicates a molecule that rotates light to the right, as in D (+) glucose, which is sometimes called dextrose. Levulose, or D (—) fructose, causes rotation in the opposite direction.

For simplicity, the structure of glucose is often shown as a straight chain but in reality the ends of the chain come together to form a ring, the —OH on one carbon atom adding across the double bond to the oxygen on the aldehyde group. This accounts for two more important types of isomerism, illustrated by the glucose molecule, which

Polysaccharides break down to disaccharides (via oligosaccharides) and then to monosaccharides by hydrolysis reactions catalyzed by enzymes. The reverse process occurs in carbohydrate metabolism when monosaccharides not broken down into carbon dioxide and water (to provide energy) combine in condensation reactions to produce polysaccharides — in animals they produce glycogen, also called "animal starch" — in which form they can be stored. The polysaccharide starch consists of a chain of numerous alpha glucose units.

Starch (polysaccharide)

Hydrolysis

Maltose (disaccharide)

Hydrolysis

Alpha glucose

Beta glucose

can either form a six- or five-membered ring. The form with five carbons and one oxygen atom in the ring is the commonest because it is more stable. Also the new —OH group formed by the addition at the aldehyde carbon atom can be either above (termed beta) or below (alpha) the plane of the ring. This difference between the beta and alpha forms is important in determining the properties of some polysaccharides.

Common carbohydrates
There are 16 stereoisomers in the glucose family (8 in the D form, with a corresponding 8 L sugars). Apart from D glucose itself, only two of these (D mannose and D galactose) are common in nature. D glucose is the sugar that occurs naturally in the blood and is used by the tissues in releasing energy. In plants, it is made by photosynthesis, whereas animals obtain their glucose either by digestion of di- and polysaccharides or from the carbon skeletons remaining after deamination of amino acids.

Fructose (fruit sugar) is common in plants, combined with glucose to make the disaccharide sucrose. It is also found in honey. Sucrose, or cane sugar, is a temporary energy store for many plants but is broken down to its constituent monosaccharide residues before it can be absorbed by animals. Maltose and lactose are two other important disaccharides. The first is a simple combination of two D glucose molecules and is found in germinating seeds such as barley and as an intermediate in the breakdown of more complex sugars. Lactose is the sugar found in milk and is formed from a molecule each of D glucose and D galactose.

As with the monosaccharides, the disaccharides are all sweet-tasting soluble solids. On the other hand starch, the common storage product of green plants such as cereals and potatoes, does not taste sweet. It is a combination of two polysaccharides, the highly branched amylopectin and the straight-chained amylose molecule, each of which consists entirely of alpha glucose residues. The mixture of polysaccharides forms microscopic granules in the storage tissues which stain blue in contact with iodine.

Glycogen is the highly branched alpha glucose storage molecule found in animals, and in some ways is so similar to starch that it is often called "animal starch." These large molecules are useful in storage because, for a given amount of material, the molecular concentration is low and does not greatly affect the osmotic pressure of the cell. Glycogen is found mainly in the muscles and liver of vertebrates. Its extensive branching facilitates the rapid mobilization from it of glucose units, when the tissues need these for energy.

Cellulose is the major constituent of the cell walls of higher plants. It is made up of beta glucose units in long helical chains. The fibrous quality of cellulose makes it useful in the textile industry as cotton. Unfortunately, its structure makes it indigestible to most animals, and those that can digest it — such as cows and termites — possess

special bacteria in their guts which contain the enzymes needed to break the beta-linkage. Chitin is a similar insoluble compound found in the shells of insects and crustaceans, although here the basic building unit is a nitrogen-containing derivative of D glucose.

Some carbohydrates occur in combination with different types of molecule. They are found in the cell membranes of mammals combined with proteins and lipids to form glycoproteins or glycolipids. Carbohydrates also perform a crucial role in the structure of the nucleic acids, which contain either the 5-carbon (pentose) sugar ribose, or its derivative, lacking one hydroxyl group, 2-dioxyribose.

Honey contains the monosaccharide fructose, which bees produce from plant nectar and store as an energy source in the comb cells of the hive.

Silkworms convert the polysaccharide cellulose — the principal constituent of the mulberry leaves on which they feed — into the silk (a protein) with which they make their cocoons.

Lipids side panel

CH₂OH — represented as CH_2OH

$$CH_2OH$$
$$CH\,OH$$
$$CH_2OH$$
Glycerol

+   +

$$3\,RCOOH$$
Fatty acids

↓   ↑

$$CH_2O.COR$$
$$CHO.COR$$
$$CH_2O.COR$$

Triglyceride (a lipid)

Lipids — in this example (above) a triglyceride — are formed by condensation reactions, esterification of glycerol and fatty acids. This is shown in the downward path (red). The reverse reaction, hydrolysis, converts fats to fatty acids and glycerol, as shown in the upward path (blue).

Phospholipids are lipids in which one fatty acid has been replaced by a phosphate group, which has then been esterified with an alcohol. This creates a complex molecule (A) which has a soluble (hydrophilic) polar end linked to insoluble (hydrophobic) fatty acid chains. When a phospholipid is spread on water, the molecules align in a layer. In theory a simple membrane could be formed by the juxtaposition of two such monolayers (B), but in practice it is unstable. Stability may be achieved by the combination of protein molecules with a phospholipid bilayer, but theories differ regarding the actual structure: one (C) proposes an arrangement of protein molecules on either side of the phospholipid layers; another (D) suggests that protein molecules are interspersed within the bilayers.

Lipids

Lipids are a large and very diverse group of compounds which occur naturally, for example as fats and oils. (The only difference between a fat and an oil is the melting point: an oil is simply a fat which is liquid at room temperature.) Lipids contain long chains or ring systems of nonpolar carbon atoms, with or without a polar group at one end. This chain makes them relatively insoluble in water but soluble in organic solvents such as ether, trichloromethane (chloroform) and benzene, and accounts for many of their biological properties.

Triglycerides

The simplest and most common lipids are the fats and oils, which are triglycerides. They are esters (a class of compounds produced by reaction between acids and alcohols with the elimination of water) made up of two kinds of unit, glycerol (a trihydric alcohol) and residues of three fatty acids. The latter are usually long-chain aliphatic acids, which are found in most but not all of those compounds classified as lipids. Triglycerides are made by esterification, a condensation reaction in which the —COOH groups of three fatty acids react with the three —OH groups on a single glycerol molecule, releasing three molecules of water. The fatty acids may be the same or different, and this factor determines the characteristics of the lipid. Many natural fats and oils like butter and olive oil are mixtures of several different triglycerides. The presence of large numbers of unsaturated fatty acids lowers the melting point of the triglyceride, as does the presence of a high pro-

portion of short-chain acids. Margarine is made solid by artificially saturating vegetable oils, which are high in unsaturated chains.

Phospholipids

Lipids are, along with carbohydrates, the main sources of energy in the diet; but they also have a number of other important functions. The most important of these is in forming cell membranes. All cells — plant and animal — take advantage of the hydrophobic (water-avoiding) properties of the fatty-acid chain. Substituting one of the fatty acids in a triglyceride with a phosphoric acid molecule gives the complex a polar end which mixes easily with water. If this phosphate residue is esterified with a polar alcohol, the product is a phospholipid. A resulting phospholipid, if spread onto water, forms a single-molecule layer on the surface, with the polar heads directed into the aqueous phase and the nonpolar fatty-acid chains directed out of it.

In principle, a cell membrane could consist of two such monolayers, with the hydrophobic chains oriented towards each other, and the polar heads facing the water on either side of this "sandwich." This stable, self-assembling configuration forms a "skin" less than a millionth of an inch thick. In practice, however, cell membranes are not simply lipid bilayers, because these would not be stable, but have various substances incorporated in them, for example glycoprotein, at the outer surface. The membrane also contains minute pores lined with protein. These pores allow substances to pass through it selectively and other proteins, intrinsic or extrinsic to the bilayer, which perform other specialized functions.

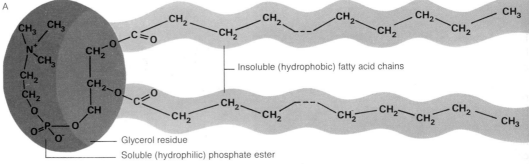

A

Insoluble (hydrophobic) fatty acid chains

Glycerol residue

Soluble (hydrophilic) phosphate ester

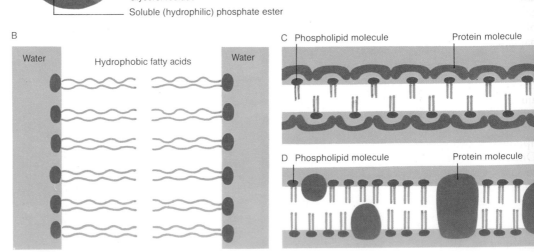

B

Water   Hydrophobic fatty acids   Water

C   Phospholipid molecule   Protein molecule

D   Phospholipid molecule   Protein molecule

Sphingolipids

Other types of lipids are also important in forming membranes. The second largest group are called sphingolipids. These are similar to phospholipids in having two long nonpolar tails and a polar head (containing phosphate), but they are not based on glycerol. Instead, they are made up of a fatty acid running parallel to, and linked by an amide bond to a long-chain molecule amino alcohol called sphingosine (or one of its derivatives). The polar group, esterified with the primary hydroxyl group of sphingosine, varies according to the class of sphingolipid, of which there are three main types. The sphingomyelins are the simplest and most common; they occur in the myelin layers around nerve cells.

The second type, cerebrosides, have no phosphorus but contain instead a carbohydrate molecule, usually a sugar. Galactose is the sugar residue in cerebrosides found in the brain, whereas glucose is the polar group in those in nonnerve tissue.

The third group of sphingolipids are called gangliosides. They possess a very large polar head containing several carbohydrate units. They are found in highest concentrations in the gray matter of the brain.

Lipids in the diet

Most lipids can be made in the body from carbohydrates or proteins but two, linoleic acid and linolenic acid, cannot be synthesized but are necessary for the maintenance of health, and so they are called essential fatty acids. Deficiency of them can result in kidney failure and retarded growth, but such conditions are rare because these acids occur in seed oils and fish, which are generally included in a balanced diet. Triglycerides make up most — around 98 per cent — of the lipids in food.

The average Western diet is high in fats. This factor is now known to be one of the causes of the high incidence of heart disease, strokes and other cardiovascular disorders in developed countries. Animal fats such as those in dairy products and red meat — which are high in saturated fats — may be partly responsible for this, by causing the build-up of abnormal fatty patches in the inner lining of artery walls. This condition, which is called atherosclerosis, can eventually lead to the blocking of vital arteries and — if the blockage happens to occur in the coronary artery — a heart attack.

Several other factors such as smoking and high blood pressure are also involved, but diets rich in animal fats are known to affect the levels of lipoproteins in the blood. Lipoproteins are complexes of lipid and protein which allow the insoluble lipid to be carried from the liver to the tissues. One type of lipoprotein, called HDLP, is thought to protect against atherosclerosis. On the other hand, increased levels of another type, LDLP, and the fatlike compound cholesterol encourage it. Changing diet, however, does not always help because heredity is thought to be the most important factor.

Palm nuts (left) are a source of vegetable oil that have provided food and fuel to humans from early times. Other common sources of vegetable oils include olives, sunflower seeds and nuts such as groundnuts and walnuts.

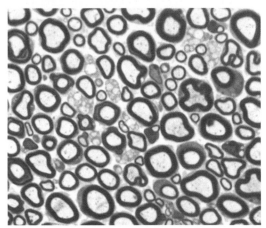

Most nerve fibers consist of axons surrounded by fat-rich medullary sheaths. The photomicrograph (left) shows nerves in cross section, with the lipids in the medullary sheaths stained to appear as dark rings around the transparent axons.

Cholesterol, a steroid, has certain properties in common with lipids but has a different structure, based on carbon rings (below).

Amino acids and proteins

Proteins are the most common large molecules in living things, and make up about 50 per cent of the total organic matter. They are important structural components of the tissues, both of cells and of the extracellular matrix. They also have an essential role in the control of the chemical reactions in the body. Enzymes and many hormones are proteins.

Amino acids

Just as carbohydrates and lipids are made from assemblies of simpler molecules, all proteins are constructed from amino acids. There are around 20 common amino acids which form a kind of alphabet from which all the words and sentences in an almost infinitely large book can be made. There is another group of amino acids, many of which are simple derivatives of the common types, which are occasionally found as part of a specific protein in a particular species. With the exception of two amino acids, methionine and cysteine (which both contain a single sulfur atom) all contain just four elements: carbon, hydrogen, oxygen and nitrogen.

It is the $-NH_2$ and $-COOH$ groups in amino acids that make them so important (at physiological pH these groups exist as $-NH_3^+$ and $-COO^-$). They can join together (with the loss of a molecule of water) to form a dipeptide, linked by an amide bond, which in this case is called a peptide bond. The process can be repeated several times to form a polypeptide, and longer chains, with perhaps more than 100 units, are referred to as proteins.

Protein amino acids are alpha amino acids, with the $-NH_2$ and $-COOH$ linked to the same carbon atom. This atom also has links to a hydrogen atom and a side chain, the nature of which defines the particular amino acid. Thus, as with monosaccharides, the carbon atom at the center of an amino acid has four different groups attached to it and is therefore asymmetric (glycine is the only exception). Because of this asymmetry, these molecules occur in D and L forms which are mirror images of each other. They also rotate the plane of polarized light — they are optically active — and the convention (+) and (−) is used to show the direction to the right or left: for example L(+) alanine. The L form is the only one found in proteins, although some D amino acids occur in antibiotics and in certain bacteria.

Structures of proteins

Proteins are very large molecules with molecular weights often in the tens of thousands. They fall into two main categories: the highly folded and roughly spherical globular proteins, such as globulin, which tend to be soluble in water, and the long fibrous proteins such as the keratin of human hair, which are insoluble. Their highly complicated shapes are determined at four separate levels.

Hemoglobin contains four heme groups, one of which is shown below. Each heme has an iron atom at its center, linked to four nitrogen atoms (pale blue). Oxygen bonds reversibly to the iron, which explains the vital role of hemoglobin in oxygen uptake and transport in blood.

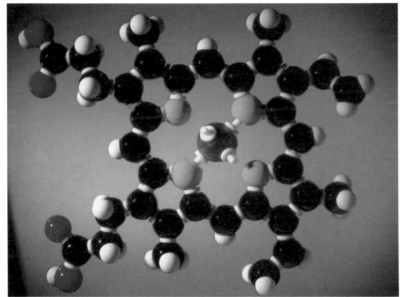

Polypeptides and proteins are formed from amino acids by a condensation reaction in which one amino acid loses $-OH$ from $-COOH$ and another loses $-H$ from $-NH_2$ to form a peptide bond. Repetition of this reaction (polymerization) converts dipeptides to polypeptides and these in turn to proteins. A standard formula for an amino acid, with the variable group R, has been used in the diagram. Breakdown of proteins to polypeptides to amino acids is the reverse process, an enzyme-catalyzed hydrolysis.

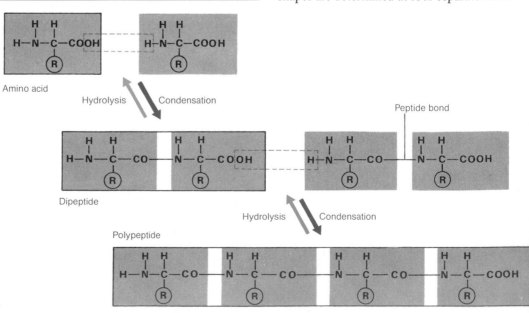

The primary structure is determined by the sequence of the amino acids, like the beads in a necklace. The sequences in proteins having the same functions tend to be similar even in very different species, but with 20 or more different amino acids to choose from, the possibilities for variation are enormous and some proteins are unique to a particular species.

The secondary structure of a protein is the result of the attraction between the peptide bonds in neighboring sections of the polypeptide chain. The C=O and N—H groups of the peptide bond are electrically bound by a weak hydrogen bond. Because these bonds can occur at regular intervals along the chain, a regular coiled structure like a telephone cord is formed, and is called the alpha helix. A different pleated form called the beta structure occurs in some proteins.

These regular structures can, however, be distorted by the influence of the polar side chains of some amino acids forming hydrogen bonds with other parts of the molecule or nonpolar side chains tending to cluster together to avoid contact with water. The sulfur atoms in the amino acid cysteine can also join together to form a strong disulfide bridge. The folding of the chains into a three-dimensional shape as a result of these breaks in the helix is called the tertiary structure.

Some proteins also possess a quaternary structure resulting from the shape taken up when two or more smaller polypeptide molecules aggregate.

Specialized proteins

Hemoglobin is an example of a protein with a quaternary structure. It is the red substance that carries oxygen in the blood. It consists of four polypeptide chains, each of which is grouped around a heme group. Each of the four hemes contains an iron atom which bonds reversibly with oxygen. Like the related compound myoglobin in muscle, hemoglobin is classed as a conjugated protein. These contain nonprotein prosthetic groups and are important in a number of biological functions. Chlorophyll, the green pigment found in plants, is closely related to heme but contains magnesium instead of iron.

Antibodies are another type of specialized protein which form part of the body's defenses against disease. They bind to a foreign protein or carbohydrate in the blood, either causing it to clump together (agglutination) or "tagging" it to trigger off a set of processes that result in its destruction. They combine with a particular point on the surface of the foreign body (called the antigen).

Proteins in food can be used by the body as a source of energy, but the body's own protein is used up only during starvation. Amino acids cannot be made from inorganic materials by humans and have to be obtained from food, although some so-called nonessential amino acids can be synthesized from other amino acids. Others, called essential amino acids, cannot be synthesized, and a deficiency in the diet will retard growth.

Protein can have great structural strength, and can be formed rapidly, as in the keratin of the antlers of a red deer, which are shed annually and have to be grown afresh every year.

Naturally occurring amino acids fall into two categories: those such as arginine, called essential amino acids, which must be absorbed from food because the body cannot synthesize them; and those such as alanine, called nonessential, which the adult body can synthesize.

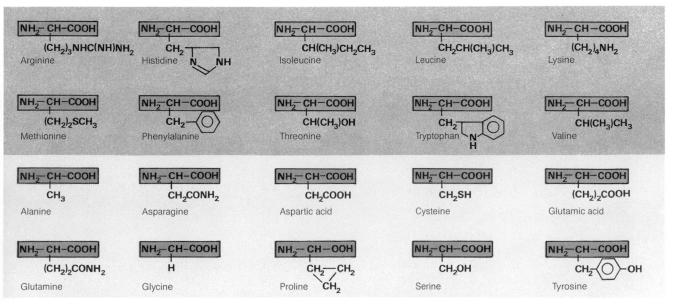

$NH_2-CH-COOH$ $(CH_2)_3NHC(NH)NH_2$ Arginine	$NH_2-CH-COOH$ CH_2 Histidine	$NH_2-CH-COOH$ $CH(CH_3)CH_2CH_3$ Isoleucine	$NH_2-CH-COOH$ $CH_2CH(CH_3)CH_3$ Leucine	$NH_2-CH-COOH$ $(CH_2)_4NH_2$ Lysine
$NH_2-CH-COOH$ $(CH_2)_2SCH_3$ Methionine	$NH_2-CH-COOH$ CH_2 Phenylalanine	$NH_2-CH-COOH$ $CH(CH_3)OH$ Threonine	$NH_2-CH-COOH$ CH_2 Tryptophan	$NH_2-CH-COOH$ $CH(CH_3)CH_3$ Valine
$NH_2-CH-COOH$ CH_3 Alanine	$NH_2-CH-COOH$ CH_2CONH_2 Asparagine	$NH_2-CH-COOH$ CH_2COOH Aspartic acid	$NH_2-CH-COOH$ CH_2SH Cysteine	$NH_2-CH-COOH$ $(CH_2)_2COOH$ Glutamic acid
$NH_2-CH-COOH$ $(CH_2)_2CONH_2$ Glutamine	$NH_2-CH-COOH$ H Glycine	$NH_2-CH-OOH$ CH_2-CH_2 CH_2 Proline	$NH_2-CH-COOH$ CH_2OH Serine	$NH_2-CH-COOH$ CH_2-OH Tyrosine

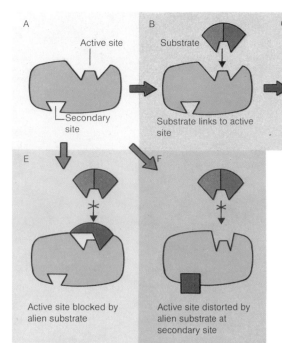

Enzymes

Enzymes are highly specialized proteins which
control the chemical reactions in all living cells.
They operate as organic catalysts, lowering the
amount of energy needed to power a reaction.
Indeed, many biochemical reactions have very
large activation energies, and most metabolic
processes would proceed far too slowly to maintain life in the absence of a catalyzing enzyme.
Most enzymes are found within cells, but some
are released to catalyze extracellular reactions,
such as the digestion of food in the stomach and
intestines.

Properties of enzymes

The essential feature of both organic and inorganic catalysts is that they participate in a reaction
by promoting a chemical change but remain unal-
tered themselves. There is an enormous variety of
enzymes because each one is usually involved in
only a single reaction. But although enzymes are
highly specific, it is interesting that the same
enzymes or groups of enzymes are often found in
a wide variety of organisms. This fact accounts for
the similarity in basic metabolic functions in
plants, animals and bacteria.

Enzymes need be present only in minute quantities to affect the rate of a reaction. They are
extremely fast-acting, and it has been estimated
that one of the most rapid enzymes, catalase
(which is found in the mammalian liver), can
break up hydrogen peroxide molecules into water
and oxygen at the rate of 40,000 molecules per
second.

All chemical reactions are reversible and the
direction of the reaction depends on the physical
and chemical conditions at a particular time. An
enzyme that breaks down a substrate A into its
products B and C is equally capable of catalyzing
the reverse reaction. It does not alter the concentrations in which the three constituents, A, B
and C, are found when the reaction reaches equilibrium, but merely reduces the time needed to
reach this state.

Because they are made of protein, enzymes
share the properties of those compounds in that
they are sensitive to temperature and acidity.
Raising the temperature generally improves the
efficiency with which an enzyme operates, but
above a certain temperature, the protein becomes
damaged. The hydrogen bonds in the molecule
start to break and the protein becomes denatured,
losing its shape and its effectiveness. Few
enzymes can work above about 113°F (45°C)
although some, such as those in the bacteria that
live in hot-water springs, can operate at higher
temperatures. Similarly, enzymes have an optimum acidity level at which they work best. This is
often around neutral (pH 7) but some, such as
pepsin (which breaks up protein in the stomach),
operate only in acid conditions, whereas others
function best in an alkaline environment.

Controlling enzyme functions

Enzymes are usually named after the reaction
they regulate. Hence an enzyme that specifically
catalyzes the removal of hydrogen from a substrate molecule is called a dehydrogenase, with the
suffix -ase signifying that the molecule is an
enzyme.

There are several theories to explain why

Many species of fungi, such as the pin mold growing on a carrot (left), obtain nutrients by secreting digestive enzymes into the foodstuff in which they are growing. The enzymes break down the large foodstuff molecules into smaller molecules, which are then absorbed by the fungus — just as human digestive enzymes break down food into small molecules that can be absorbed through the wall of the alimentary canal.

enzymes are specific and how they operate, but they all center on the enzymes' three-dimensional structure. The simplest thory, the "lock-and-key" hypothesis, postulates that a substrate molecule (the "key") attaches itself to an active site (the "lock") of an enzyme molecule, forming a temporary complex. The active site has a particular shape, and so only a substrate with the complementary shape can attach itself to this site — just as a lock only accepts a key with the right shape. Hence molecules with different shapes cannot attach to the active site. Also, if the active site is distorted by excessive heat, the substrate molecule itself will no longer fit.

However, some other compounds may be close enough in shape to the substrate that they fit into the active site of the enzyme. These alien substrates are not changed through contact with the enzyme but compete with the true substrate for active sites, thereby inhibiting the enzyme's activity.

There are two types of inhibition: competitive, in which an alien molecule forms only a temporary bond with an enzyme, and noncompetitive, in which an alien inhibitor molecule either permanently blocks an enzyme's active site or affects it by temporarily binding to a site elsewhere on the enzyme. Many inhibitors, particularly of the noncompetitive type, act as poisons; cyanide — which binds with an enzyme necessary for cellular respiration — is a good example.

More complicated theories of enzyme action hold that an enzyme exists in two separate shapes. The most energetically stable shape may be different when the substrate is present from when it is not. Some enzymes do not work without the presence of what are called cofactors. Some cofactors participate in the enzyme reaction. Others probably lock into the enzyme away from the active site but hold it in the correct position to receive the substrate.

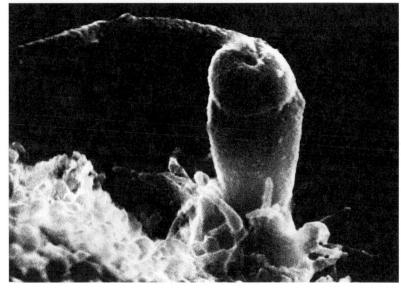

Enzyme systems

Most enzymes work as part of a chain of reactions in a metabolic pathway. The product of one enzyme-induced reaction then becomes the substrate for another. To prevent the wasteful production of unnecessary amounts of a particular substance, the whole reaction sequence is controlled by the slowest step. The activity of enzymes is often inhibited by the presence of large amounts of the product of the metabolic pathway.

Enzymes may consist entirely of protein, or of protein linked to a prosthetic group which helps to maintain the shape of the molecule, or participates in the reaction (or both). Trace metal minerals (such as iron and cobalt) in the diet are often necessary for use in these prosthetic groups, and water-soluble vitamins are often important because they are cofactors in an enzyme system.

Sexual reproduction in many animals depends on enzymes. When the head of a sperm hits the outer membrane of an egg, a small sac of special enzymes at the front of the sperm's head breaks open. The released enzymes break down the egg's membrane, so allowing the sperm to penetrate and fertilize the egg — as shown in the scanning electron micrograph (above). After the egg has been fertilized, its outer membrane undergoes little-understood changes that make it resistant to the enzymes of subsequent sperm, thereby preventing multiple fertilization.

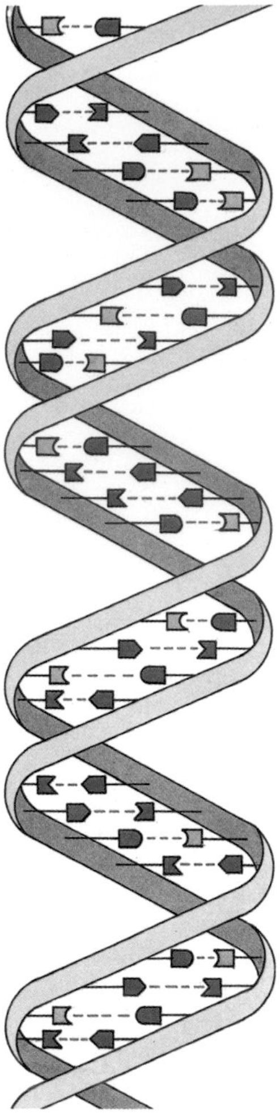

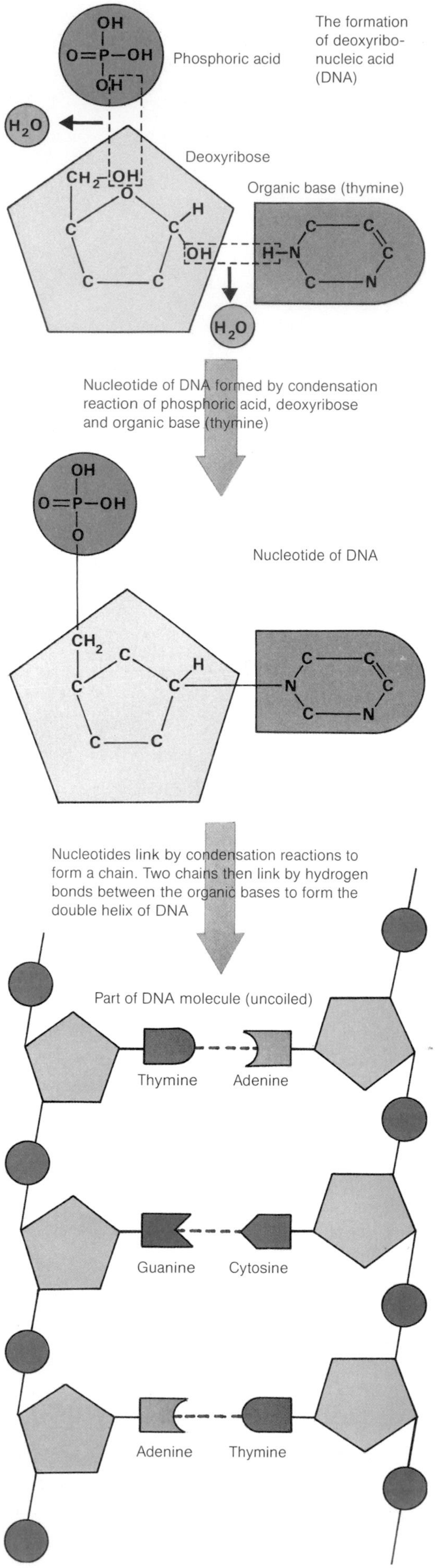

The formation of deoxyribonucleic acid (DNA)

Phosphoric acid

H_2O

Deoxyribose

Organic base (thymine)

H_2O

Nucleotide of DNA formed by condensation reaction of phosphoric acid, deoxyribose and organic base (thymine)

Nucleotide of DNA

Nucleotides link by condensation reactions to form a chain. Two chains then link by hydrogen bonds between the organic bases to form the double helix of DNA

Part of DNA molecule (uncoiled)

Thymine — Adenine

Guanine — Cytosine

Adenine — Thymine

DNA is a giant nucleic acid molecule consisting of two nucleotide chains twisted into a double helix (above). The sequence of reactions by which it is formed is shown in the diagram (right). The first stage is the synthesis of individual nucleotides by a condensation reaction between phosphoric acid, the five-carbon sugar deoxyribose and an organic base (thymine, adenine, guanine or cytosine — thymine is the example used in the diagram; for clarity the base and sugar molecules have been represented schematically). Many nucleotides then link by further condensation reactions to form a nucleotide chain. Finally, two nucleotide chains link by forming hydrogen bonds (dotted blue lines in the diagram) between their organic bases, producing the twisted, ladder-like DNA molecule.

Nucleic acids

Nucleic acids are the means by which information about the structure and function of a living organism is stored and passed on to the next generation. They consist of only two types of molecules, deoxyribonucleic acid (DNA) and ribonucleic acid (RNA) and they are found in the cells of all living organisms from viruses to man.

Structures of nucleic acids

Both RNA and DNA are made up of recurring units called nucleotides. These consist of complexes of three different molecules, a five-carbon monosaccharide (sugar), an organic base and phosphoric acid. In RNA the sugar is ribose, whereas in DNA it is deoxyribose, a ribose derivative in which one hydroxyl group is replaced by a hydrogen atom.

The organic bases of DNA are the two pyrimidine compounds cytosine (C) and thymine (T), and the purine molecules adenine (A) and guanine (G). In RNA, uracil (U) is substituted for thymine. The sugar molecule is attached through a condensation reaction to both the phosphoric acid and the base, with the phosphoric acid residue linking to the sugar of the next nucleotide. Hence any sugar molecule in the middle of the chain is linked to one base and two phosphoric acid residues.

Although the basic constituents of nucleic acids had been known for many years, it was not until 1953 that Francis Crick and James Watson at Cambridge University worked out the three-dimensional structure of the DNA molecule. They suggested that the bases of two nucleotide chains are connected together by hydrogen bonds, with the sugar and phosphate running alternately along each side like a ladder. The ladder is twisted into a regular helical formation, the famous DNA double helix. The purine and pyrimidine bases are always found in complementary pairs, with adenine linking with thymine and guanine in combination with cytosine.

It is in the DNA double helix that all the information about the structural proteins and enzymes which make up the organism is stored. A few viruses contain only RNA and do not possess DNA. But in all other species the purpose of the RNA is to transcribe the information stored in the DNA and transfer it to sites in the cell (the surfaces of ribosomes, which themselves tend to be attached to a system of membranes in the cytoplasm known as the endoplasmic reticulum) where it is translated in the making of protein.

The genetic code

The sequence of organic bases in a DNA molecule forms what amounts to a four-letter code which must provide the words in an enormous encyclopedia of possible protein types. As there are at least 20 words or amino acids in proteins, a single base obviously does not give sufficient information to specify what is needed to make the protein.

Three bases together give a choice of 64 (4 x 4 x 4) combinations, and 61 of those possible triplets of the four bases A, C, G and T code for specific amino acids. Several different combinations therefore code for the same amino acid. The remaining three perform the same function as the full stop at the end of a sentence, showing that the last amino acid in the protein has been reached. This theory that three bases code for a particular amino acid is supported by experimental evidence; the three-base unit is referred to as a codon.

DNA is found mainly in the nucleus of plant and animal cells, whereas proteins are manufactured outside the nucleus, by organelles called ribosomes within the cytoplasm. A complex chain of events links the two.

Protein synthesis

The double helix of DNA is the largest molecule in the cell, but RNA exists as much smaller molecules and in several different types. To relay information to the ribosome, the two strands of the DNA double helix must first split apart, like a zip, along the appropriate part of the molecule. A molecule of messenger RNA (m-RNA) is then formed from free nucleotides by pairing with the bases of the section of DNA coding for the required protein. As the RNA bases pair only with the complementary bases of the DNA, the information is coded "in negative" and the sequence on the RNA must be transcribed back into its original form. This is done after the m-RNA has moved out of the cell and taken up position on the ribosome. Another RNA molecule called transfer RNA (t-RNA) picks up a free amino acid and takes it to the ribosome. The enzymes that control this attachment are highly specific, and each molecule of t-RNA carries only one type of amino acid.

The t-RNA molecule is smaller than m-RNA, consisting of a single nucleotide chain twisted back on itself into a cloverleaf-shape. At one end is a sequence of three bases which attach to the appropriate complementary codon on the m-RNA. The amino acid at the other end is enzymatically joined to the polypeptide chain as the m-RNA slides along the ribosome. Several protein molecules may be formed simultaneously from the same m-RNA molecule.

The same genetic material is found in all the cells of an organism, but not all the cells produce the same proteins and there are also differences in the rate of production between each cell. The mechanism by which the function of a gene (the section of DNA that codes for a particular protein) is controlled is poorly understood. But production of a protein can be stopped or slowed down in three ways: by the DNA stopping making m-RNA, by preventing attachment on the ribosome, or by increasing the rate at which m-RNA is destroyed.

Mutations

DNA is a huge, extremely complicated molecule and it is inevitable that mistakes in its replication occasionally occur. These are called mutations and happen when the wrong base is coded for, or sections of DNA are removed or put in the wrong place. Ionizing radiation and some chemicals increase the rate at which these mistakes occur, probably by inhibiting the natural repair mechanisms. Some mutations produce inheritable diseases, usually where the change causes the production of the wrong amino acid, which renders an enzyme ineffective by altering its shape.

Mutations are now also known to cause cancer, which is an abnormal growth of particular cells. But not all mutations are damaging. Some cause beneficial variation in a species, which is an important mechanism in evolution.

Protein synthesis (below) begins in the cell nucleus (A) with the slitting apart of a section of DNA and the subsequent synthesis of a messenger RNA (m-RNA) molecule. The m-RNA is formed by the bases of free nucleotides pairing with complementary bases of the DNA (the bases that occur in RNA are adenine, guanine, cytosine and uracil, this last substituting for the thymine that is found in DNA). The m-RNA then moves through a pore in the nuclear membrane and becomes attached to a ribosome (B). Next, transfer RNA (t-RNA) molecules transport amino acids to the ribosome. In addition to an amino acid at one end, each t-RNA has a sequence of three bases at the other; these bases attach to the complementary three-base sequence (called a codon) on the m-RNA. As the m-RNA moves along the ribosome, the amino acid on the t-RNA links to an adjacent amino acid, thereby building up the polypeptide chain. This process is repeated until the protein molecule coded for by the m-RNA is complete.

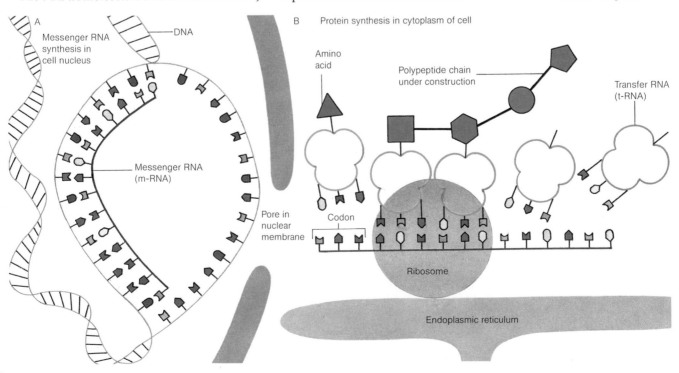

A Messenger RNA synthesis in cell nucleus — DNA

Messenger RNA (m-RNA)

Pore in nuclear membrane

B Protein synthesis in cytoplasm of cell

Amino acid

Polypeptide chain under construction

Transfer RNA (t-RNA)

Codon

Ribosome

Endoplasmic reticulum

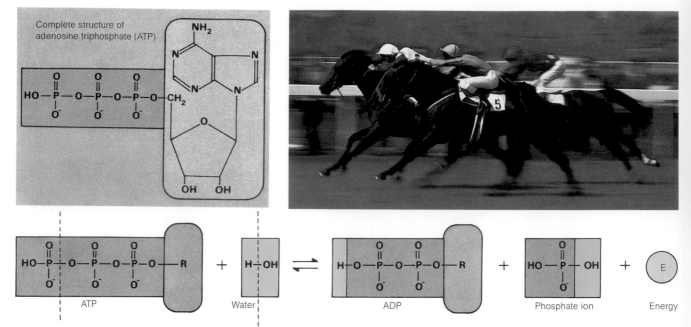

Complete structure of adenosine triphosphate (ATP)

ATP + Water ⇌ ADP + Phosphate ion + Energy

During strenuous activity large amounts of energy are required quickly, especially by the muscles. The immediate source of this energy is adenosine triphosphate, ATP (structure shown top left). The energy is released when ATP reacts with water to form ADP (adenosine diphosphate) and a phosphate ion (above). The reaction is reversible, so ADP and a phosphate ion can be reconverted to ATP by supplying energy.

Biochemical energy

Energy is needed for nearly all the vital processes that take place in animals and plants. All the energy used by animals is ultimately derived from plants eaten as food. But plants and many bacteria are able to trap light (or occasionally chemical energy) and use it to convert inorganic materials, such as carbon dioxide and water, into complex organic products. This energy, "stored" in carbohydrates and lipids, can then be used to do work in the organism. This may be the chemical work of the biochemical processes, electrical work in nerve cells, or mechanical work in the muscles. The amount of energy needed obviously depends on the size and complexity of the organism. But the levels of physiological and physical activity are also important. For example, the energy requirements of a sleeping human being may increase tenfold or more during strenuous exercise.

Anabolism and catabolism

Metabolism, the network of biochemical reactions which underlie living processes, consists of constructive (anabolic) and destructive (catabolic) pathways. The breaking down of large molecules into smaller units releases energy which may be used to build other large units. Many thousands of these reactions are going on in the body all the time. The pathways for particular types of molecules, such as proteins or carbohydrates, do not remain separate but converge so that energy can be released from any available fuel. Similarly, in anabolic processes, a particular compound in excess can be converted to a different material for use in growth or for storage. This flexibility ensures that anabolic and catabolic processes are in balance in a normal organism.

The role of ATP

Adenosine triphosphate (ATP) is a relatively simple compound derived from the purine base adenine. It consists of adenine and the sugar ribose (together constituting the nucleoside adenosine) linked to a phosphate group to which, in turn, are linked two more phosphate groups through pyrophosphate bonds. ATP is the readily accessible energy "currency" used by all plants, animals and bacteria. The terminal phosphate group can be broken off (so ATP becomes ADP — adenosine

The hydrogen and electron carrier system associated with the breakdown of glucose is illustrated in the diagram below; details of this system are explained in the main text.

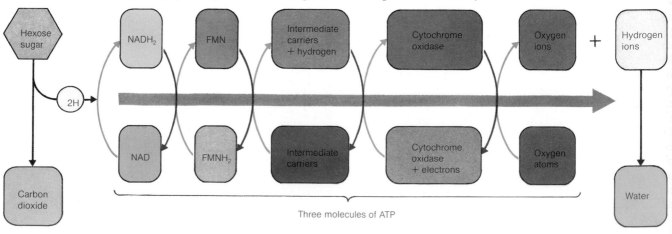

Three molecules of ATP

diphosphate) to make energy available for immediate use — for electrical or mechanical work, for example, or to form chemical bonds in new molecules, which may thus store some of the energy released from the ATP. Alternatively, the energy released in other reactions can be used to convert ADP back to ATP. Hence ATP provides "ready cash" for the organism to spend or save in its "bank account" of carbohydrates and lipids. Energy is also stored in proteins; this is called upon to some extent during fasting and can become the only energy source during starvation, when the other resources are exhausted. Some other compounds also contain energy-transferring phosphate bonds, particularly creatine phosphate, which provides an emergency reserve to regenerate ATP rapidly in contracting muscle. In terms of an organism's total energy metabolism, however, these compounds are of only minor importance.

ATP is produced from ADP by a highly complex sequence of reactions which is essentially the same in all organisms. The breakdown of ATP releases energy, and so the phosphorylation of ADP must involve the input of the same quantity of energy. The energy is produced by the gradual oxidation of molecules such as hexose sugars and fatty acids, although these may have been made originally by breaking down other molecules. Eventually the molecules used to produce ATP are oxidized to carbon dioxide and water, just as they would be if burnt directly in air. For this reason, cellular respiration is often referred to as "burning" food, although the description is slightly misleading. Direct oxidation involves the production of heat energy which is of only limited use to a cell, and in excess would be damaging. Instead, cell respiration involves the gradual stripping away of hydrogen atoms from the foodstuff and the controlled transfer of this hydrogen to oxygen so that the production of water is coupled with the release of small "packets" of chemical energy. In this way, the conversion is not only more controlled than it would be in a single-step reaction, it is also a more efficient source of useful energy.

Hydrogen and electron carriers

The production of ATP is accompanied by a sequence of reactions in which the hydrogen atoms shed by the foodstuff are passed between compounds known as hydrogen carriers. In fact, some of the compounds do not accept hydrogen atoms but only electrons from the hydrogens (resulting in the release of hydrogen ions). These compounds are called electron carriers and they contain iron or copper atoms, which change oxidation state as electrons are taken up or passed on. Hydrogen and electron carriers — found in the cell organelles called mitochondria — are first chemically reduced by receiving two hydrogen atoms or electrons before being oxidized back to their original form in passing the atoms or electrons on to the next carrier.

The first carrier is called nicotinamide adenine dinucleotide (NAD), derived from the vitamin nicotinic acid. This is reduced to $NADH_2$ by hydrogen from foodstuff before passing the hydrogen to the next carrier, a derivative of vitamin B_2 (riboflavin) called flavine mononucleotide (FMN). This in turn is reduced to $FMNH_2$ before passing on to the next carrier in a long and complex chain that includes a substituted benzoquinone called ubiquinone and a set of colored compounds called cytochromes. The final set of carriers is in the enzyme cytochrome oxidase, which transfers electrons to oxygen atoms; the resultant oxygen ions then combine with hydrogen ions taken up from the medium to form water. The entire respiratory chain is built into the structure of the membranes of the mitochondria and is subdivided into three spans between $NADH_2$ and oxygen. As two hydrogens or electrons cross each span, the energy released is used to make a molecule of ATP; hence the complete oxidation of $NADH_2$ yields three molecules of ATP.

Glycolysis

The first phase in the breakdown of glucose is called glycolysis, and in the absence of oxygen is the only method by which most organisms can obtain energy. It takes place in the cytoplasm of the cell and results in the formation of a three-carbon molecule called pyruvic acid. If oxygen is available, the pyruvic acid is moved into the mitochondria where the next stage occurs. But without oxygen, glycolysis can keep going only if pyruvic acid is continuously converted into lactic acid (in animals) or ethanol (some other organisms). Lactic acid (2-hydroxypropionic acid) is the substance that accumulates in muscles and causes fatigue when the oxygen is used up during strenuous exercise. Both lactic acid and ethanol are potentially toxic, and are formed only as a temporary measure until the oxygen is replenished. In contrast, some bacteria called anaerobes

Some types of bacteria and algae are unusual in the way they obtain energy for their metabolic processes, utilizing inorganic sources such as hydrogen sulfide or iron salts. For example, the rust colored area in the foreground (below) contains large numbers of iron bacteria (which are responsible for the color) that derive their energy by oxidizing iron (II) compounds to iron (III) compounds.

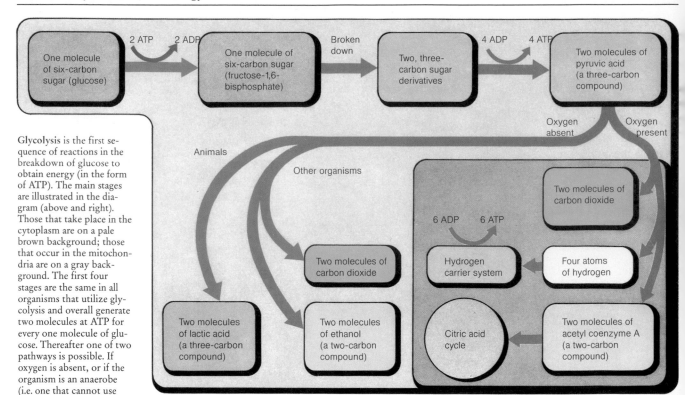

Glycolysis is the first sequence of reactions in the breakdown of glucose to obtain energy (in the form of ATP). The main stages are illustrated in the diagram (above and right). Those that take place in the cytoplasm are on a pale brown background; those that occur in the mitochondria are on a gray background. The first four stages are the same in all organisms that utilize glycolysis and overall generate two molecules at ATP for every one molecule of glucose. Thereafter one of two pathways is possible. If oxygen is absent, or if the organism is an anaerobe (i.e. one that cannot use oxygen), the pyruvic acid is converted to either lactic acid or ethanol, with no further generation of ATP. If, however, oxygen is present, the pyruvic acid is broken down — with the production of six molecules of ATP from the one original glucose molecule — to acetyl coenzyme A (acetyl CoA). The acetyl CoA then enters the citric acid (or Krebs) cycle, which results in yet more ATP being produced. The principal stages in this cycle (which can also produce energy from fats and proteins) are illustrated in the diagram on the opposite page; the details are described in the main text.

use the method as the source of all their energy and for them oxygen is toxic.

The method is an inefficient means of converting the energy in glucose into useful packets of ATP energy, because there is a net gain of only two ATPs per glucose molecule degraded. It begins with the input of two ATPs, the first used to convert the sugar into glucose-6-phosphate, the second to form fructose-1,6-bisphosphate. This is then split into two three-carbon triose sugar derivatives, which are converted to pyruvic acid with the manufacture of four ATPs. Hydrogen atoms are also produced and are taken up by NAD, but without oxygen these do not pass along the hydrogen carrier system and the NAD can be regenerated to maintain glycolysis only by transferring the hydrogen from $NADH_2$ to the pyruvic acid, thus forming lactic acid.

Acetyl coenzyme A

Before oxidation can proceed to the second stage, called the Krebs or citric acid cycle, the pyruvic acid must lose another carbon atom (as carbon dioxide). This reaction produces two hydrogen atoms which pass along the hydrogen-carrier chain, thus yielding three molecules of ATP and one molecule of acetyl coenzyme A (acetyl-CoA). This latter substance is extremely important, not only in the breakdown of carbohydrates but also in the oxidation of fats and proteins. It is formed by linking the remaining two-carbon molecule of acetic (ethanoic) acid with coenzyme A, a derivative of the B-group vitamin pantothenic acid. Just as the coenzyme NAD may act as a carrier for the transfer of hydrogen atoms, CoA is a carrier for acetyl groups in a number of reactions.

Krebs or citric acid cycle

The Krebs cycle is named after the Anglo-German biochemist Hans Krebs, who discovered it in 1937. It is the final stage in aerobic oxidation and the most significant in terms of total energy yield. An equivalent of 12 ATP molecules are formed for each acetyl group oxidized via the cycle, and overall 38 molecules are formed during the complete oxidation (through aerobic glycolysis and the citric acid cycle) of a single glucose molecule. This compares with only two ATP units formed in anaerobic glycolysis in which a large amount of the total glucose energy remains locked up in the lactic acid or ethanol produced. The cycle is a highly complex sequence of reactions catalyzed by a number of enzymes, all found within the mitochondria. It begins with the transfer of the acetyl group of acetyl-CoA to the four-carbon oxaloacetic acid molecule to produce a six-carbon molecule of citric acid.

The citric acid is rearranged to produce isocitric acid, and then two carbon atoms are sequentially lost (as carbon dioxide) to form first alpha-oxoglutaric acid and then succinic acid. In both these reactions, hydrogen atoms are transferred to the coenzyme NAD, with the production of three ATP molecules as the resultant molecules of $NADH_2$ pass their hydrogen atoms to oxygen down the mitochondrial respiratory chain. In the second reaction a molecule of guanosine triphosphate (GTP) is made (like ATP, this is an energy-transferring molecule derived from a purine base, guanine). The GTP subsequently transfers its terminal phosphate to ADP, so making ATP.

During its conversion to fumaric acid, the succinic acid molecule loses two hydrogen atoms, used in hydrogenating FAD, with the formation of another two ATP molecules as $FADH_2$ is oxidized by the respiratory chain. Although no more carbon atoms are lost after succinic acid, two further conversions take place. A water molecule is added to form L-malic acid and another pair of hydrogen atoms are lost to produce oxaloacetic acid and another $NADH_2$, the oxidation of which leads to the formation of another three ATP molecules. The newly-formed oxaloacetic acid is then ready to start the cycle again.

Fatty acid oxidation

As well as being structural components of the cell membranes and various complex functional compounds (such as hormones), lipids are stored as droplets of triglycerides (fats) in adipose tissues. These triglycerides are a valuable source of energy. They are particularly important sources for various internal organs such as the heart, and for prolonged activity in skeletal muscle as, for example, in migrating birds; fats are also key energy sources for hibernating mammals.

As in carbohydrate oxidation, the final oxidation of fats takes place in the mitochondria. But first the fat must be mobilized. It is enzymatically broken down in the adipose tissue to release free fatty acids, which are transported in the blood to the cells of the energy-requiring tissue. In the cells, but before entering the mitochondria, the fatty acids are linked with coenzyme A (CoA). The resultant fatty acyl-CoA then enters the mitochondria via a complex process that involves the intervention of another molecule, carnitine.

The conversion of fatty acids to CoA derivatives requires the input of energy as ATP, but this investment is soon recovered. The reaction takes place in a repetition of a four-step sequence in a process called beta-oxidation, or the fatty acid spiral. The net result of each sequence in the spiral is removal of the two terminal carbon atoms on the fatty acid as acetyl-CoA, leaving a fatty acyl-CoA, two carbons shorter, ready for another sequence in the spiral. The acetyl-CoA joins the citric acid cycle and the process is repeated until the final fragment left is itself acetyl-CoA. In each sequence of the fatty acid spiral, four hydrogen atoms are removed; two of these are taken up by FAD (flavin adenine dinucleotide) and two by NAD. Oxidation of the resultant $FADH_2$ and $NADH_2$ yields five molecules of ATP.

The process is varied slightly if the acid contains an odd number of carbon atoms (when the final fragment is propionyl-CoA) or is partly unsaturated (containing carbon-carbon double bonds). But in all cases it is a very good energy source: not only are five ATP molecules produced for each two-carbon fragment, but each fragment is itself oxidized to form 12 more ATPs. In the breakdown of the 16-carbon palmitic acid, for example, a total of 129 ATP molecules are produced.

Protein oxidation

The adaptability of the oxidation pathways described earlier is especially important in the

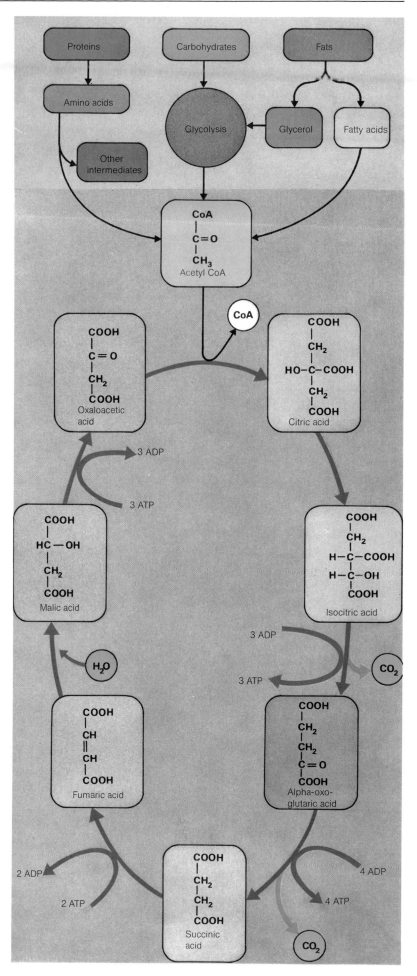

breakdown of the third energy source, protein. The structures of the component amino acids vary to the extent that the preliminary pathway for each is different. Some are converted into acetyl-CoA, whereas others are converted into intermediates that join at later stages of the citric acid cycle. In each case, the amino group on the amino acid has to be removed first. In normal circumstances, protein plays a minor role in total energy production. Except during starvation, it provides (mainly after conversion to glucose) only 10 per cent of the body's needs.

Amino acid degradation and the urea cycle

As well as being oxidized occasionally to produce energy, amino acids are also broken down during the normal turnover of protein in the tissues. If the amino acid unit is not needed in forming new protein, it is oxidized, as are excess amounts absorbed in food. The amino group removed at

the start of amino acid oxidation has to be expelled from the body because ammonia is very poisonous. The citric acid cycle intermediate alpha-oxoglutarate accepts the amino group and then releases it as an ammonium ion. (Another intermediate, oxaloacetate, also readily accepts amino groups, forming aspartate, which plays an important part in nitrogen elimination.) Ammonia is toxic at very low concentrations and so it has to be converted to a form in which it can be passed out of the body in small volumes of water. This compound is urea, and the urea cycle pathway in which it is made occurs in most land animals. Fishes and some other aquatic animals excrete ammonia directly because they have plenty of water available and are not likely to dehydrate. Before entering the cycle, the ammonia reacts with bicarbonate and ATP to form carbamoyl phosphate, which then reacts with ornithine to produce citrulline. This compound combines with aspartic acid to form arginosuccinic acid which, in the next stage, breaks down into the by-product fumaric acid and arginine. Finally, arginine breaks down to ornithine (for reuse in the cycle) and urea.

Photosynthesis

Photosynthesis is the means by which plants, and ultimately animals, derive the energy they need for their metabolic processes. Light energy is used to synthesize organic molecules from water and carbon dioxide. In some ways it is therefore a reverse of the process of respiration. It takes place in organelles of the plant cell called chloroplasts, which have structural and functional analogies to mitochondria. Chloroplasts are thought to have originated from symbiotic blue-green algae, mitochondria from symbiotic aerobic bacteria. Unlike mitochondria, however, which are found in all organisms, chloroplasts occur only in green

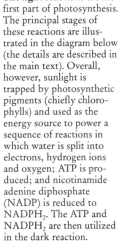

The light reactions are the first part of photosynthesis. The principal stages of these reactions are illustrated in the diagram below (the details are described in the main text). Overall, however, sunlight is trapped by photosynthetic pigments (chiefly chlorophylls) and used as the energy source to power a sequence of reactions in which water is split into electrons, hydrogen ions and oxygen; ATP is produced; and nicotinamide adenine diphosphate (NADP) is reduced to $NADPH_2$. The ATP and $NADPH_2$ are then utilized in the dark reaction.

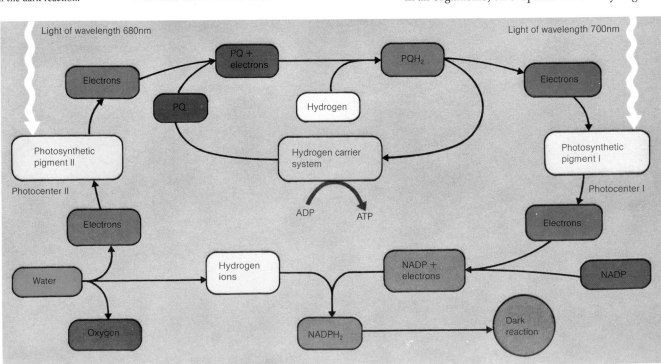

algae and higher plants. Certain bacteria and the blue-green algae also photosynthesize, but the process takes place in structures that are simpler than chloroplasts. It has been estimated that more than 10^{20} joules of light energy is fixed by plants each year during photosynthesis, with much of this occurring in marine algae.

The light reaction

Photosynthesis occurs in two stages, the first of which needs the presence of light whereas the second can take place in darkness. The light reaction is dependent on the green plant pigment chlorophyll, a porphyrin compound containing magnesium (and related to hemoglobin and the respiratory cytochrome molecules). There are several similar pigments in chlorophyll although only the blue-green chlorophyll (a) is common to all plants. Most plants also contain yellow-green chlorophyll (b), xanthophyll (yellow), carotene (orange) and pheophytin, a gray pigment which is probably a breakdown product of chlorophyll. Each pigment absorbs light of a slightly different wavelength. The purpose of the light reaction is to release hydrogen from water for the reduction of carbon dioxide to carbohydrate. It takes place in structures called thylakoids, which resemble stacked plates. All the later reactions occur in the liquid matrix, or stroma, of the chloroplast.

There are two types of light centers in thylakoid membranes, called photosystems I and II. Absorption of light with a wavelength of about 680nm by pigment at center II excites the molecule so that electrons are released, reducing a quinone (plastoquinone, PQ) on the other side of the membrane. The electrons are replaced by removing them from water molecules, releasing oxygen gas and hydrogen ions. Absorption of light at around 700nm by chlorophyll at center I also causes electron release, this time to reduce the hydrogen carrier nicotinamide adenine dinucleotide phosphate (NADP) on the other side of the membrane. Because PQ and NADP are both hydrogen carriers, their reduction by electrons produced in the light reactions also requires the uptake of hydrogen ions from the external medium. PQH_2 is then reoxidized by a sequence of carriers (including cytochromes) similar to those of the mitochondrial respiratory chain. Electrons are eventually transferred to replace those ejected from center I, with hydrogen ions being released inside the thylakoid. The net effect of the two photosystems is that NADP is reduced to $NADPH_2$, water is split, oxygen is released and hydrogen ions are separated across the membrane. These ions then return, but utilizing a special enzyme that conserves the energy released by this process as ATP — the same principle that underlies ATP production by mitochondria.

The dark reaction

The dark reaction begins with the combination of a carbon dioxide (CO_2) molecule with the five-carbon sugar ribulose-1,5-bisphosphate. This splits to form two molecules of 3-phosphoglyceric acid. It is then reduced by the hydrogen carried by $NADPH_2$ to form a three-carbon triose phosphate. This can then be used to build up the hexose sugars which are later stored as starch.

But not all the triose phosphate goes to make sugars, because some has to be used in reforming more ribulose bisphosphate so that carbon dioxide may continue to be taken up by this pathway, known as the Calvin cycle. Some of the steps of this cycle, including the reduction of phosphoglyceric acid, require ATP, but this (like $NADPH_2$) is available as a result of the light reactions. Studies with radioactive carbon have shown that phosphoglyceric acid is also vital in building up amino acids and fats.

The dark reaction (below) is the second part of photosynthesis and is the process in which energy (in the form of ATP) and $NADPH_2$ generated by the light reactions are used to build up complex organic molecules such as starch, fats and proteins. The various stages of the dark reaction are explained in detail in the main text.

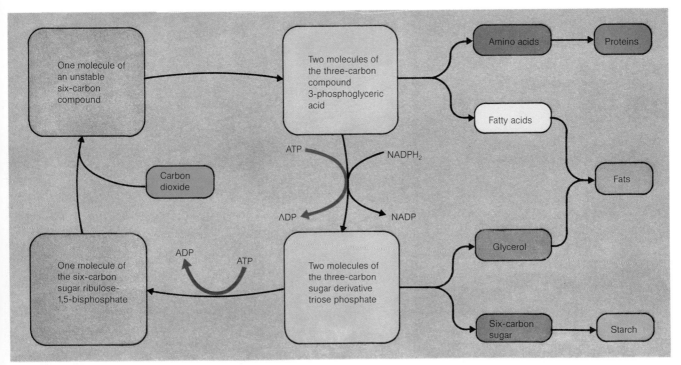

The metamorphic stages of insects — such as the Gulf fritillary butterfly emerging from its pupal case (above) — are controlled by two hormones: juvenile hormone, which keeps the insect in its larval stages, and ecdysone, which causes an adult to develop when the level of juvenile hormone is low.

Steroids have structures based on cholesterol, as exemplified by the molecular diagrams below of testosterone, estradiol (an estrogen), progesterone (a progestin), norethynodrel (a synthetic progestin) and cortisone.

Biochemical messengers

Hormones are chemical messengers which, when made in minute quantities in one part of a multicellular organism, are transported to a target cell in another part, where they stimulate a particular biochemical or physiological activity. In vertebrates, they are formed in endocrine, or ductless, glands which empty directly into the bloodstream. Hormones are also known to occur in some invertebrates, such as insects, and similar compounds — called phytohormones — control activities such as growth and flowering in plants.

Hormones are produced in such small quantities that they have proved difficult to isolate for studies on their structure and function. But modern techniques are rapidly increasing knowledge of their action on the target cells and the relationship between hormonal control and the other main method of communication in an organism, the nervous system. They are generally stored in an inactive form in the cells of the endocrine glands, and are released following the arrival of appropriate nerve impulses.

A hormone is carried in the blood to a receptor on the target cell. The interaction of the hormone with its receptor may then cause the release of a second messenger, which either modulates an enzyme pathway or controls the genes which code for that enzyme. The compound cyclic adenosine monophosphate (c-AMP) is involved in several different hormone activities as the second messenger. It is formed from adenosine triphosphate (ATP) by the enzyme adenylate cyclase and modulates a variety of enzymes which produce the cellular response (for example, the mobilization of glycogen to release glucose from the liver following stimulation by the hormone glucagon).

Types of hormones

There are three main classes of hormones in vertebrates. The peptide hormones include all those produced in the brain by the hypothalamus and pituitary glands, together with two hormones — insulin and glucagon — made in the pancreas. They are proteins containing between three and about 200 amino acid residues, and are frequently active in much smaller concentrations than are other hormones. A deficiency of a particular hormone often results in disease. Somatotropin, or anterior pituitary growth hormone, is an example; its absence causes dwarfism or retarded growth, whereas an excess produces acromegaly, which involves excessive growth on the hands, feet and facial bones of adults.

The second main class contains the best understood of all hormones, epinephrine. These are amine compounds produced in the central (medulla) part of the adrenal glands and are derivatives of the amino acid tyrosine. Epinephrine (also called adrenalin) is closely related to the nerve transmitter substance norepinephrine, and demonstrates again the interrelationship between hormonal and neural control. It prepares the body for an emergency and, in common with many other hormones, acts on several different target organs:

Cholesterol CH_3 CH_3 HO

Testosterone CH_3 OH   H CH_3 O

Estradiol CH_3 OH   H   HO

Progesterone CH_3 $COCH_3$ H CH_3 O

Norethynodrel CH_3 OH   C:CH   O

Cortisone   O CH_3 $COCH_2OH$ OH CH_3 O

it increases the heart rate and blood pressure, releases glucose for energy from the liver, and speeds up the formation of ATP in the muscles.

The steroid hormones — the third main class — are produced in several parts of the body, including the gonads and the outer layer (cortex) of the adrenal glands. Steroids are fatty compounds based on cholesterol and, because of their insolubility in water, are transported in the blood joined to protein molecules.

The structures of steroid hormones are very similar but their biological effects are extremely diverse. Many are concerned with sexual functions — examples are estrogens produced by the ovaries in females and testosterone by the testes in males. They also have other effects on the body. The steroid cortisone is used medically to suppress the immune system and male androgen hormones are used (illegally) by athletes to increase muscle size. Vitamin D is also a steroid and, following chemical modification in the body, acts as a hormone controlling the synthesis of a protein concerned with the uptake of calcium.

Pheromones are another group of hormones with a specific function outside the body. They are scents used to attract members of the opposite sex during courtship and to ward off rivals of the same sex. They are particularly important in insects, with some moths able to detect the presence of a mate at a distance of several miles. There is evidence to show they are also important in higher animals.

Vitamins are a diverse group of compounds whose only common characteristic is that they cannot be synthesized by an organism and are therefore essential ingredients of the diet. They are needed only in minute quantities and, in the case of the water-soluble ones, usually perform some function as part of an enzyme pathway. Some vitamins, such as vitamin D in humans, can be produced in small amounts but the rest of the requirement has to come in food. Other organisms may not be able to produce a vitamin directly but obtain it by a symbiotic relationship with another organism, such as bacteria in the gut.

Vitamins are usually classed according to their solubility in water (the B complex and vitamin C) or in fat (vitamins A, D, E and K). Although these compounds are needed continuously because of the breakdown of existing vitamins in the body, it is possible to store certain vitamins in the liver. Excessive amounts of the fat-soluble vitamins can be toxic, but a deficiency is more common. Scurvy (insufficient vitamin C), beriberi (lack of B_1) and rickets (D) are all deficiency diseases.

Antibiotics

Antibiotics are another class of chemicals made to create a response in another organism. They are produced by microorganisms and prevent the growth of, or even destroy, rival bacteria. They are extremely important in medicine — penicillin produced by the penicillium mold is the most famous example.

Epinephrine (structure right) is secreted in large amounts when the body has to cope with unusual physical demands or is presented with a potentially hazardous situation, or both (as with the skier jumping, below). The effect of the hormone is to prepare the body for "flight or fight": by increasing the heart rate, raising blood pressure, stimulating glucose release and accelerating ATP formation.

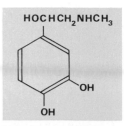

The tips of plant shoots respond to light by growing toward it — as demonstrated by these barley shoots (left); the middle shoot is bent towards the right (the direction of the light source) but the other two are not because their tips are covered by foil caps. This response to light occurs as a result of hormones (called auxins) secreted by the tips of the shoots.

Breadmaking, like brewing, is an ancient biotechnical process. Both use yeast fermentation, resulting in the production of carbon dioxide (which makes the bread dough rise).

Biotechnology

For thousands of years, people have exploited the biochemical activities of living cells. Bread- and beermaking, for example, depend upon the ability of microscopic yeast cells to convert sugars to carbon dioxide and ethanol. Cheese- and yogurtmaking also depend on the activities of single-celled life forms, the microorganisms.

Every microorganism is a tiny factory, capable of carrying out the basic biochemical functions described earlier in this section. Human understanding of biochemistry has grown dramatically in the past few decades. This explosion of science is now providing the basis for a new industry — biotechnology.

Industrial enzymes

Simple examples include some of the uses now made of enzymes. Yeast fermentation occurs because of a series of enzymes present in yeast

Genetic engineering involves modifying an organism's chromosomes and thus the genes they carry. The above, fruit fly, (Drosophila sp.) is normal, whereas the lower one is a dwarf mutant whose genes were altered by irradiation.

cells, although the catalytic activity of an enzyme does not depend on its being inside a cell. Many enzymes produced naturally by microorganisms are extracellular. They may, for example, be released from the cell into the surrounding medium in order to break down some of it into molecules which the cell can absorb as nutrient.

Many different enzymes have now been isolated as pure chemicals. Some are produced on a very large scale, usually in a relatively crude form, and used as industrial materials. Washing powders today often contain "biologically active" ingredients. These are enzymes intended to improve the efficiency of washing by biodegrading insoluble materials, such as protein, which make up part of the dirt on clothes.

Enzymes are also used in various ways in food manufacture. One method of making a chocolate with a liquid center is to start with a solid center and add an enzyme which breaks down part of the solid content to a liquid after the chocolate surrounding it has hardened.

Genetic engineering

The area of biotechnology which has attracted most attention in the past decade is that known as "genetic engineering." Partly as a result of developments in enzyme biochemistry, it is now possible to perform a variety of operations on the nucleic acid polymers which form the genetic inheritance of a cell.

The nucleic acid contains a number of different genes strung along it. Each of these codes for the manufacture of a particular protein. It is now possible to cut the nucleic acid polymer and splice in extra genes not normally found in the microorganism. Provided that conditions can be found to make the organism "express" this foreign gene, large quantities of specific proteins can be made using fermentation processes similar to those which have been used for many years to produce antibiotics.

An example in which the technique can be applied usefully is in the manufacture of insulin. This peptide hormone has to be taken regularly by large numbers of diabetics, whose bodies are unable to maufacture it naturally. Normally, insulin is produced in special cells in the pancreas of mammals. The insulin supplied to diabetics is obtained mostly from pig pancreases by a process of extraction and purification. The purification involves many stages, in order to remove other proteins that could be harmful if injected into human beings.

Microorganisms do not normally produce insulin. But the insulin gene has been inserted into bacteria and these have then been cultured to produce insulin. Although most diabetics have no adverse reaction to pig insulin, some do — possibly because its structure differs slightly from that of human insulin. Insulin made by genetic manipulation of bacteria is chemically identical with human insulin and therefore overcomes the problem of intolerance. A further advantage is that culturing of microorganisms is easier to control as an industrial process than is producing insulin from animal glands, where problems of continuity and quality of the raw material can arise.

Large-scale biotechnology

Products such as insulin are of high value and are therefore most likely to recoup the very large costs of research and development involved in establishing a genetically engineered process. Nevertheless, there have been some attempts to introduce biotechnology processes for lower-value products, which generally need to be produced on a larger scale.

Protein is an important dietary ingredient and processes are now available to produce tonnage quantities of microbial cells as feedstuffs for farm animals. The amount of space such processes require is many times less than that needed to produce the same amount of protein from green plants; for example, about 80,000 tons (70,000 metric tons) per year of bacterial protein can be produced in a factory covering only a few acres or hectares and using methanol (made from natural gas) and ammonia as its raw materials.

Over the next few years, it is likely that an increasing number of high-value products will be produced biotechnologically. These will come not only from microorganisms, but also from large-scale fermentation of both animal and plant cells. Until recently, it has been difficult to maintain cultures of cells from multicellular organisms in a healthy state for long periods, but modern developments in biochemical techniques have now made this possible. New vaccines and important, complex organic molecules could be produced in these ways. In the longer term, biotechnology holds out the hope of producing commodity materials, such as plastics, by fermentation directly from living organisms, thus freeing mankind from dependence on diminishing supplies of oil.

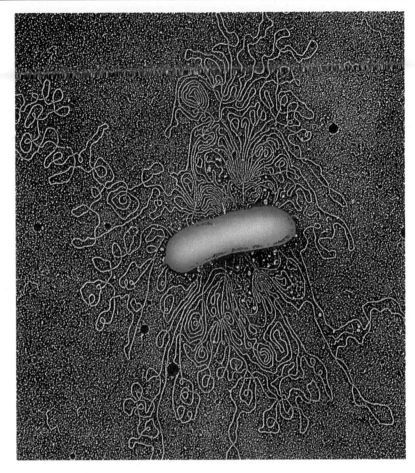

Strands of DNA emerge from the bacterium *Escherichia coli* — the most frequently used organism for genetic manipulation — shown above magnified approximately 40,000 times.

Single-cell protein as food for humans or animals can be made by growing bacteria or yeast on various organic materials such as oil, natural gas (methane) and even wood pulp. The process described (below, left) uses bagasse — the waste cellulose from sugar cane — and a bacterium called *Cellulomonas*.

The simple process (below) employs a yeast, *Candida* sp., and hydrocarbons from an oil refinery to make protein animal feed.

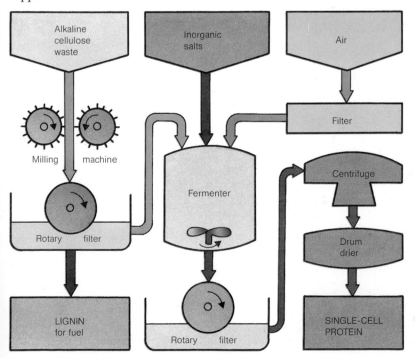

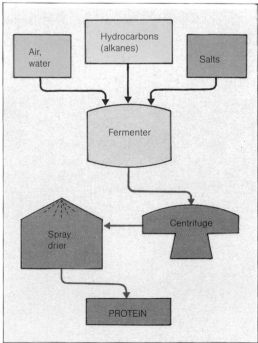

Analytical chemistry

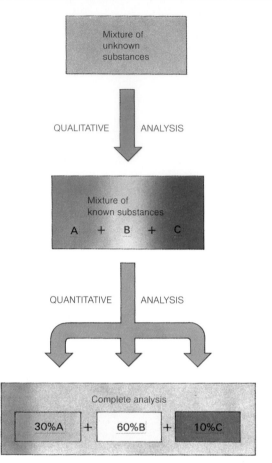

Mixture of unknown substances

QUALITATIVE ANALYSIS

Mixture of known substances

A + B + C

QUANTITATIVE ANALYSIS

Complete analysis

30%A + 60%B + 10%C

Chemists use two main types of analytical techniques. Qualitative analysis reveals which elements or compounds are present in a substance or mixture of substances. Quantitative analysis determines how much of each component there is. Certain techniques, such as some forms of spectrographic and chromatographic analysis, can make both determinations simultaneously, telling the chemist both what is there and in what amounts.

century, it was shown that Gadolin's yttrium was a mixture of several elements. First, two other elements — erbium and terbium — were isolated from it. Then these two were found to be mixtures of several elements, one of which was subsequently called ytterbium, thus celebrating Ytterby for the fourth time.

Separation

Analytical techniques are methods for finding out something specific about an element or compound. They are based on differences in the chemical and physical properties of materials. Sometimes, these differences are very small, making analytical work difficult and often time consuming. An important set of analytical techniques are those which separate different compounds or elements. It was refinements in such techniques which led, for example, to the isolation and identification of the full range of rare earth elements.

Very sophisticated separation techniques have been developed during the twentieth century, although often — as in the case of chromatographic methods — they have roots in nineteenth-century research. It is now possible to separate very small amounts of complex molecules from one another. Much of the progress made in recent years in understanding the chemistry of biological processes, for example, has depended on the development of such techniques.

Detection

Analysis does not always depend on separation, however. One of the triumphs of modern analytical chemistry has been the development of techniques that can detect very small quantities of a particular substance in a complex mixture. Such techniques may be qualitative or quantitative; that is, they may reveal only what material is there (the "quality") or they may also be able to tell how much of it is present (the "quantity"). Thus, it is possible to detect very low concentrations of particular pollutants in the atmosphere. In general terms, the atmosphere is a simple mixture of nitrogen and oxygen gases. But these make up only about 99 per cent of its total volume; the remaining 1 per cent is a complex mixture which makes quantitative analysis of trace constituents a demanding task. It can also be an essential one, for the build-up of some pollutants could have a serious effect on world climate, making life as we know it more difficult than it is already.

The techniques that have been developed for measuring very small quantities of particular molecules often depend on their physical rather than their chemical behavior. This means, in general, how they behave in the presence of different types of radiation — infrared, radio waves, or laser light, for example.

Analytical chemistry is often regarded as one of the least exciting parts of modern chemistry. Yet much of the science is built on results obtained from analyses, and many of chemistry's important uses — whether in helping to solve a murder case or discovering whether a river is polluted — rely on analytical techniques.

The primary aim of analysis is to find out how a material is constituted. Thus, a chemist may analyze a mixture to find which compounds are present, or a pure compound to find from which elements it is formed. All of the structures of chemical compounds illustrated in the previous sections of this book were originally worked out from various types of analysis.

The discovery of the different chemical elements also depended largely on developments in analytical techniques. For example, four elements — all relatively uncommon — derive their names from the Swedish town of Ytterby. Towards the end of the eighteenth century, the Finnish chemist Johann Gadolin identified what he thought was a new element in a mineral (later called gadolinite) which he found near Ytterby. The element was named yttrium. It is one of a group of elements called the rare earths (lanthanides), which have very similar chemical properties. As analytical techniques improved during the nineteenth

High-altitude balloons (left) are used for taking gas samples from the upper atmosphere. By analyzing the samples, chemists can monitor the composition of the air at various heights — checking, for example, on the concentrations of ozone and carbon dioxide. They can also detect the presence of any pollutants.

Methods of separation are important in noninstrumental chemical analysis. Solvent extraction methods (right) are often used in organic analysis. Components in a mixture of substances are dissolved out selectively in turn by different solvents. Liquids of different boiling points can be separated by distillation; the mixture is heated, and the component liquids distilled off one at a time at about their boiling points.

One of the problems facing chemists in recent years is that the ability to measure very small amounts of different substances has outstripped the ability to understand their significance. Thus, it is now possible to measure the presence of certain impurities in food at almost unbelievably small concentrations. What is not understood, in many cases, is whether or not there is a threshold level below which such an impurity will exert no harmful effect and is therefore of no significance. This, however, is not really a problem for the analytical chemist, whose basic task is to find ways of obtaining the information.

Rare talent

To be a good analytical chemist is a rare and valuable talent. It requires a clear understanding of the enormous advances in chemistry which come as a result of new techniques. It is perhaps significant that a large number of recent Nobel Prizewinners have been associated with advances in analytical chemistry. In recent times, only two scientists have won two Nobel Prizes for scientific work and one, Frederick Sanger, did so for pioneering work in chemical analysis, first in the study of proteins and second in the analysis of nucleic acids.

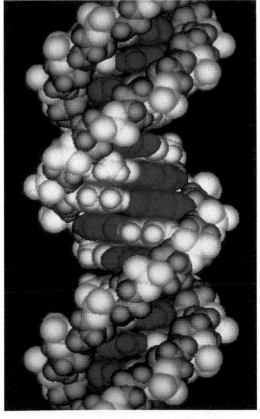

The model of the double helix structure of a molecule of DNA (deoxyribonucleic acid) shown here (left) was "drawn" by a computer. Laboratory computers have become an essential part of modern instrumental analysis, either to store and compare data or to present results as print-outs, alphanumeric displays on a screen, or colored computer graphics.

Classical analysis

Despite the considerable growth of instrumental methods of chemical analysis, many analyses are still carried out by what are termed "classical methods" using techniques and procedures that have proved to be reliable and reproducible over many years. Such methods are of particular value for use in laboratories which lack the more advanced instruments. Classical methods fall into two main groups: qualitative analysis is used to find out what is present; and quantitative analysis deals with the procedures for determining how much of a substance is present.

Qualitative inorganic analysis
The identification of cations (positive ions) and anions (negative ions) is based upon the characteristic reactions that each species undergoes in solution. This means that in most instances the first important step in the analysis is dissolving

any solid samples by treatment with water, acids or bases. Once a suitable solution had been prepared, identification of the component ions can be carried out by well established chemical reactions.

The process of identification depends upon three main types of reactions. The first is the formation of colored precipitates by treatment with chemical reagents under specified conditions of acidity or alkalinity. The second is the development of characteristic colors in solution, usually as a result of using selective and highly specialized reagents. And the third is the evolution of easily identified gases from the solution following reaction with acids, alkalis or selective reagents.

Tests of this nature are frequently referred to as "spot tests" and may be carried out on small volumes of the sample solution placed in slight depressions on a glazed white tile. The tile surface provides an ideal background for viewing any color changes occurring in the solutions. Identification of any individual ion, however, depends on obtaining positive reactions with a range of several reagents. A single positive reaction can serve only as an indication to identification but is not by itself conclusive; further confirmation is always necessary.

Qualitative organic analysis
The first steps in the identification of organic compounds are based upon physical tests, including appearance, color, odor and solubility, followed by a determination of either the melting point (of solid substances) or the boiling point (of liquids). By comparing these results with those in published tables, the analyst can rapidly narrow down the number of chemical possibilities.

Further identification then follows a well-established procedure of determining the elements present in the compound (by decomposing it into inorganic substances), and establishing whether it is an acidic, alkaline or neutral substance.

This process is followed by a series of tests to determine the nature of the reactive groups in the compound. The tests are carried out on very small quantities of material and are based upon the characteristic reactions of the hydroxyl, carbonyl, amino, halogen and carboxylic acid groups likely

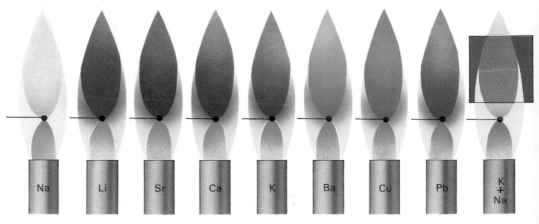

Flame tests can reveal the presence of various metals. A sample (preferably a chloride) is burned in a Bunsen flame on a platinum wire. The yellow flame of even traces of sodium (Na) masks the others but can be blocked by blue glass (A). Na = sodium; Li = lithium; Sr = strontium; Ca = calcium; K = potassium; Ba = barium; Cu = copper; Pb = lead.

to be present. Many of these tests give rise to colored precipitates or solutions when a positive result is obtained. Once the substance has been provisionally identified, it may well be confirmed by carrying out a further well-established reaction to form a chemical derivative which should be characteristic for the analyzed compound. Sometimes further confirmation of the identity is carried out using spectroscopic methods.

Quantitative inorganic analysis

In order to ascertain how much of a compound, element or ion is present in a substance two main classical procedures are available. These are referred to as gravimetry and titrimetry.

Most gravimetric determinations are concerned with converting the ion or element being studied into a pure stable compound that can be readily weighed. Gravimetry thus involves the three main steps of dissolving the weighed sample to give a solution, precipitating out the required element or ion in a new pure chemical form using a very selective reagent, and filtering, drying and weighing the precipitate. By calculation, the percentage of the required ion or element present in the starting material can be established.

Such determinations are usually carried out on as little as a thirtieth of an ounce (less than 1 gram) of substance and are capable of giving an accuracy of better than 1 per cent. The selectivity of the precipitation reactions is often achieved by using one of a wide range of organic ions. In some gravimetric determinations the precipitated solid is further reacted or strongly heated to provide a stable chemical compound suitable for weighing.

In titrimetry all determinations are carried out in solution and precipitation of solids rarely has any part to play. The technique is based on reacting carefully measured volumes of very pure chemical reagents with similar volumes made from the materials being studied. For instance, if the acid content of a fruit juice needs to be determined, a measured quantity of the juice is reacted with a steadily increasing volume of an appropriate base until all the acid has reacted. From a knowledge of the chemical reaction that has taken place and the strength of the solution of the base, it is possible to calculate how much acid was present in the sample of fruit juice.

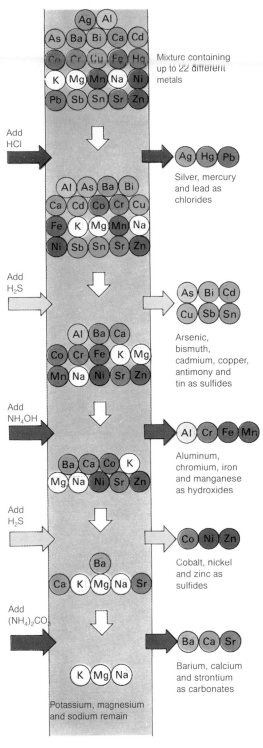

Mixture containing up to 22 different metals

Add HCl

Silver, mercury and lead as chlorides

Add H₂S

Arsenic, bismuth, cadmium, copper, antimony and tin as sulfides

Add NH₄OH

Aluminum, chromium, iron and manganese as hydroxides

Add H₂S

Cobalt, nickel and zinc as sulfides

Add (NH₄)₂CO₃

Barium, calcium and strontium as carbonates

Potassium, magnesium and sodium remain

Qualitative inorganic analysis using "wet chemistry," on a macro or micro scale, depends on precipitating groups of metals from a mixture as insoluble compounds. The diagram (left) outlines the principle of the method for a mixture of 22 common metals (which would normally be in the form of their salts or other simple compounds). A sample is dissolved or suspended in water, and dilute hydrochloric acid (HCl) added. Any silver, mercury or lead present is precipitated as the insoluble chloride. Hydrogen sulfide (H₂S) gas is bubbled through the remaining solution to precipitate the next group of metals as their sulfides. The solution is made alkaline with ammonium hydroxide (NH₄OH) and yet a further group precipitated as hydroxides. Then H₂S is again bubbled in, precipitating another group of sulfides (this time from alkaline, not acid, solution). Finally ammonium carbonate, (NH₄)₂CO₃, is added, to precipitate barium, calcium and strontium as their carbonates. The only metals of the original mixture left in solution at this stage are potassium, magnesium and sodium. The individual metals in each group are identified by specific tests. Some salts, such as phosphates, interfere with the method and have first to be removed, if they are present.

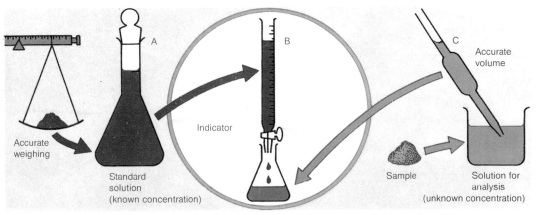

In volumetric analysis, a standard solution (of known concentration, prepared by accurate weighing and dissolution in a volumetric flask, A) is titrated from a burette (B) against a known volume — measured using a pipette (C) — of a sample of unknown concentration. The end point, when the reaction between sample and titer is complete, is denoted by an indicator.

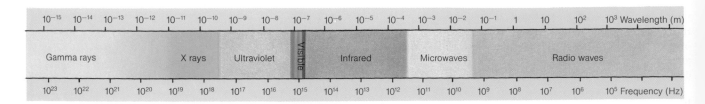

Gamma rays   X rays   Ultraviolet   Visible   Infrared   Microwaves   Radio waves

| 10^{23} | 10^{22} | 10^{21} | 10^{20} | 10^{19} | 10^{18} | 10^{17} | 10^{16} | 10^{15} | 10^{14} | 10^{13} | 10^{12} | 10^{11} | 10^{10} | 10^{9} | 10^{8} | 10^{7} | 10^{6} | 10^{5} Frequency (Hz) |

Modern methods of spectroscopy between them employ the whole range of the electromagnetic spectrum (above), from short-wavelength gamma rays and X rays to the long wavelength radio waves used in nuclear magnetic resonance techniques.

Spectroscopic analysis

Among the most useful scientific instruments for chemical analysis are those involving the radiation and absorption of energy from different regions of the electromagnetic spectrum. By using different wavelengths of energy, it is possible to excite selectively either the electrons in bonds of molecules, electrons in various atomic orbits, or even the nuclei in the different atoms. The molecules and atoms of any single substance give the same characteristic patterns for the absorption of radiation. These spectra can be used as "fingerprints" to identify unknown substances and often to determine how much of that substance is present.

Atomic absorption and emission spectroscopy
The region of the electromagnetic spectrum from part of the ultraviolet, through all of the

visible light to part of the infrared portion of the spectrum is particularly useful for identifying elements present in materials and for quantitative measurements. When atoms absorb energy from this part of the spectrum, electrons "jump" from one orbit to another of higher energy level. A similar amount of energy is released when the electrons move back to their original, lower-energy orbits. As a result, elements can be identified by the characteristic series of wavelengths — seen as spectral lines — that each produces when these energy changes occur. The best-known series of emission lines are those in the visible spectrum first examined by the Swiss scientist Johann Balmer (1825—1898). The atoms of the elements studied are excited by an electric spark or arc, and the various wavelengths of light emitted are separated by a large prism and used to expose a photographic plate. Identification of the elements present is carried out by comparing the wavelengths of the lines with those produced from known pure substances.

Quantitative measurements using emission spectroscopy are best carried out by using an argon plasma torch. The sample in solution is fed into a tail-shaped plasma flame formed from argon gas under the influence of a strong magnetic field produced by a radio-frequency generator. The production of the characteristic emission lines can be easily measured using this system.

Most quantitative measurements may, however, be more readily carried out using atomic absorption spectrometry. In this technique, measured amounts of solution or sample are burnt or vaporized to produce free atoms, which can absorb the energy emitted from a special lamp

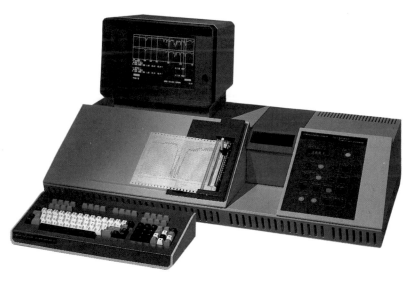

An infrared spectrometer (above) records the characteristic wavelengths at which various groupings in a molecule absorb radiation, and can therefore be used to identify the molecule. The absorption is usually expressed in terms of percentage transmission, so that low transmission corresponds to strong absorption. The spectrum of aminopentane (right) has strong absorption bands for NH_2, the C—H bond and the C—N bond.

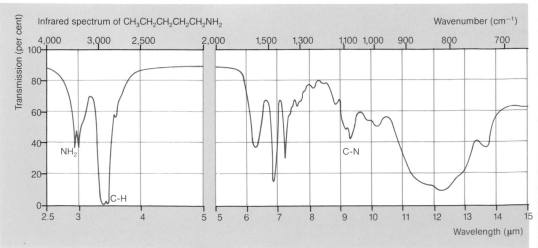

Infrared spectrum of $CH_3CH_2CH_2CH_2CH_2NH_2$ Wavenumber (cm^{-1})

which produces only the lines that are characteristic for the element under investigation. The extent of the energy absorption is proportional to the number of atoms present and hence to the concentration of the element in the sample.

Electronic spectra

Organic compounds also give rise to absorption spectra in the ultraviolet, visible light and infrared regions, but these are not discrete lines. They occur as very broad bands, sometimes in a series of peaks and troughs in and around the visible region. These spectra arise due to the absorption of energy by electrons forming the molecular bonds, and are characteristic for each compound. Although they can be used to assist in identification, it is in the realm of quantitative analysis that they are of greatest value because the main absorptions are obtained with only small quantities of material and can be measured very accurately. As such they are specially useful for determining minute amounts of colored and unsaturated compounds.

Infrared and Raman spectroscopy

Just as a tuning fork vibrates at specific frequencies, so the bonds between atoms absorb energy and vibrate when irradiated with radiation at wavelengths that lie in the infrared region of the spectrum. The actual wavelength at which a vibration occurs is characteristic for the two elements forming the bond. This means that in any molecule formed from several different atoms, there are many different values at which vibrations occur, and an infrared spectrum covering this region consists of a series of fairly sharp peaks of varying intensities. The infrared spectrum of a compound is specific to that compound, and is used as a "fingerprint" for identification purposes.

Raman spectroscopy also studies the vibrations of interatomic bonds, but is used for structures that do not necessarily give straightforward infrared spectra. Unlike infrared spectroscopy, it can be readily applied to samples dissolved in water and the information obtained is complementary to that from infrared. It is especially useful for locating functional groups in molecules, as well as for quantitative analysis of mixtures of complex organic compounds.

Raman spectra occur when intense monochromatic visible light is passed through samples, the energy changes in the oscillations of the bonds being shown by a series of emission peaks at wavelengths on each side of the irradiating wavelength. In modern Raman spectroscopy, lasers are used as the excitation source.

Resonance spectroscopies

In recent years a number of advanced spectroscopic techniques have been developed which involve subjecting molecules to strong magnetic or radio-frequency fields. These include nuclear magnetic resonance (NMR), electron spin resonance (ESR), and Mössbauer spectroscopies.

NMR produces a spectrum of a series of absorption peaks corresponding to energy changes of the nuclei of one element (hydrogen atoms are most frequently studied) in a compound. The various elements in different chemical groups absorb at different magnetic field strengths, so that compounds can be identified from the spectral patterns produced.

In ESR it is the spin energy levels of the electrons of substances such as rare earths and transition metal ions which are studied when under the influence of magnetic fields. For analytical purposes ESR is of particular value in investigating free radicals which have been formed by breaking a two-electron bond so that each portion, or radical, is left with one unpaired electron. Mössbauer spectroscopy is a study of nuclear energy levels by the absorption or emission of gamma radiation using a moving energy source.

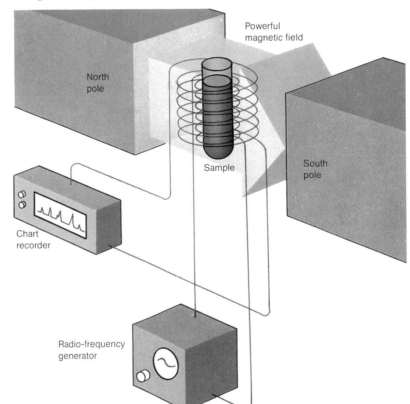

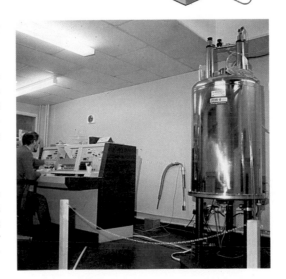

In a nuclear magnetic resonance spectrometer, a coil (shown green, above) detects the absorption of high-frequency radio waves (supplied by the blue coil) by hydrogen nuclei in a sample in a gradually varying powerful magnetic field. A modern instrument (left) uses a liquid helium cooled superconducting magnet (inside the stainless steel container), chained off to remind people of the presence of an extremely powerful magnetic field.

Advanced instrumental analysis

Recent developments in electronics and new materials have led to a number of advanced analytical instruments and procedures becoming established in scientific laboratories around the world. Such instruments can be very expensive, but the information they can provide about chemical structures is of enormous value.

Mass spectrometry

If chemical compounds are bombarded in a vacuum with streams of electrons or highly concentrated light beams from a laser, the molecules lose electrons and form positively-charged ions (cations); some molecules are broken down into smaller fragments which also carry positive charges. The overall result is that each molecular species gives rise to a series of positively-charged ions that are characteristic of the original compound and can be used for its identification.

Mass spectrometry is the name given to the study of these ions, which may be separated from each other and measured in an instrument called a mass spectrometer. There are various forms of mass spectrometer, the most common consisting of a powerful magnetic field into which the fragments of the bombarded molecules are directed. Within the field the different ions follow individual paths related to their mass, and by varying the strength of the magnetic field they are brought to focus separately on an electron multiplier detec-

tor. The mass spectrum of a compound consists of a series of peaks that differ in height, depending on the relative numbers of ions of different mass values. On each spectrum the highest peak usually corresponds to the mass of the positive ion of the compound itself, the rest of the spectrum being peaks obtained from progressively smaller and smaller fragments.

Compounds can be identified from their mass spectra by the shape and distribution of the various peaks. The fragmentation of similar molecules follows a pattern, and some chemical groupings — such as hydroxyl, carboxylic acid, and amino — are more readily detached than others. The loss of such distinctive parts of the molecules can also be an aid to identification. The value of mass spectrometry has been greatly enhanced by linking it with gas chromatography and with computerized storage of reference data.

Surface analysis

A variety of instrumental procedures have been developed for investigating the surfaces of materials or thin films deposited on other solids. They are of considerable importance for the detailed and accurate study of catalysts, oxidation of surfaces, and rates of corrosion. All of the methods depend on the impact of various forms of energy on the surfaces, and measurements of particles or energy released by the beam. The activating beam may consist of electromagnetic radiation, ions, electrons or X rays, depending upon the depth of surface to be examined. Low energy electrons, for example, penetrate only a few nanometers, whereas X rays are used for greater depth studies. Such techniques have made it possible to examine surfaces to a depth of more than 1 micron (one thousandth of a millimeter).

These methods of investigation are used to show how surface layers of materials differ from the layers underneath, and to reveal the lack of uniform distribution of metals in alloys and

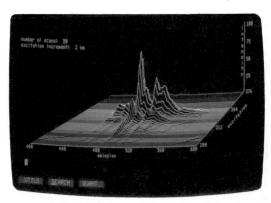

A spectrophotometer (left) measures concentrations by comparing the optical density of a liquid sample with that of a solution of known concentration.

More recent computerized machines can display information as "three-dimensional" colored graphics (above).

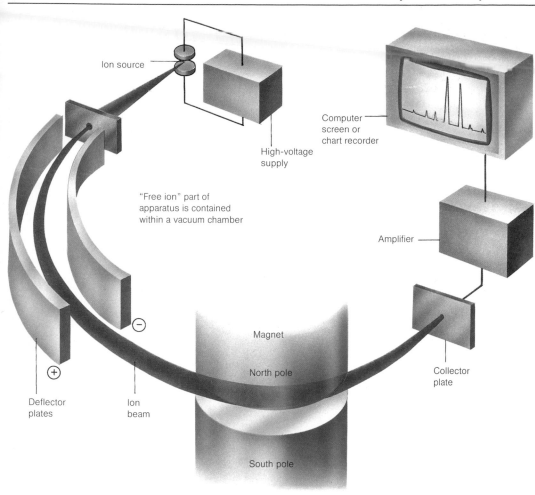

Ion source

High-voltage supply

"Free ion" part of apparatus is contained within a vacuum chamber

Computer screen or chart recorder

Amplifier

Magnet

North pole

South pole

Collector plate

Deflector plates

Ion beam

⊖

⊕

In a focusing mass spectrograph, a beam of positively-charged ions (cations) is deflected by curved electrically-charged plates, and further deflected and focused by a magnetic field. Ions of different masses come to a focus in slightly different places and can be recorded on a photographic plate or, as here, collected as an electric current which is amplified and plotted by a chart recorder or video screen. The mass spectrogram produced takes the form of a bar chart (as shown below), which relates the intensity — that is, abundance — of particular ions to their mass-to-charge ratio (m/e).

sprayed coatings. In X-ray fluorescence, for instance, samples are irradiated with an X-ray beam and individual elements in the sample are determined by the characteristic lines of energy they emit. Auger spectroscopy, on the other hand, measures the characteristic energy of electrons released from atoms that have been ionized as a result of electron bombardment.

Radiochemical analysis

The use of radioactive isotopes has been of great value in the development of analyses for low concentrations of materials. A radioactive isotope is chemically identical with any corresponding stable isotope of the same element, and so it undergoes all the same chemical processes and can at all times be followed quantitatively by means of its radioactive emissions. Often only traces of the radioactive isotopes are needed, and they can be incorporated into chemical compounds to produce "labeled" substances which can be used and followed throughout in precipitation reactions, solvent extractions and separations. A measurement of the radioactivity at any stage in a process establishes how efficient that step has been by the proportion of radioactivity carried through to the next step.

Isotope dilution analysis is a special application of radioactivity in which a known weight of a radioactively labeled compound is added to a mixture containing an unknown amount of the

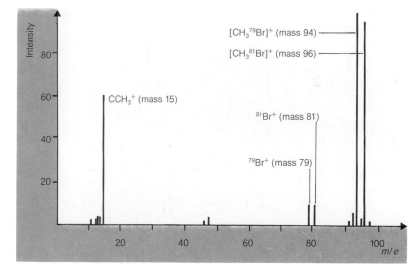

same — but unlabeled — compound. After complete mixing a small amount of the mixture is removed and treated selectively for the particular chemical compound. By measuring the degree of dilution of the radiotracer, it is possible to calculate the amount of nonradioactive compound.

Activation analysis is different in that a nonradioactive sample is irradiated with high-energy neutrons to convert some of the inactive atoms into radioactive isotopes. By measuring the amount of radioactivity produced under defined conditions, it is possible to calculate how much of that element was originally present.

The mass spectrogram of bromomethane (CH_3Br, above) reveals that the bromine in the compound exists as a nearly equal mixture of two isotopes of atomic masses 79 and 81.

Thermal analysis

Modern thermal analysis includes various techniques in which changes in some physical property of a material (such as its weight) are measured as its temperature is increased. The measurements are often carried out on very small samples, and can be used to give an idea of how substances will change over extended periods under normal conditions of temperature.

Thermogravimetry
This is the best-known form of thermal analysis. It consists of measuring the weight change of a sample as the temperature is steadily increased,

and the resulting weight/temperature graph is characteristic for the material being studied. Changes in weight usually occur sharply at specific temperatures and correspond to the breaking of chemical or physical bonds. They are often associated with the loss of volatile substances such as water, carbon dioxide or oxygen from the molecules of the sample.

A thermal gravimetric graph provides a large amount of information from the sizes and shapes of the changes in the line. Horizontal sections of the graph indicate no weight change and hence no decomposition, whereas slopes and curves show that a weight change has taken place due to some material loss. The curve is also quantitative in the sense that weight losses between one plateau and another correspond to losses of definite parts of the molecular structure.

The measurement of thermogravimetry is carried out on a thermal balance, in which the sample is suspended from a balance arm in a special crucible in a quartz sleeve within the heated metal furnace block. As the weight of the sample drops during the heating, which may go as high as about 3,000°F (about 1,650°C), the balance beam is maintained in a fixed horizontal position by a turning force produced by a current-carrying coil in a magnetic field, which is increased (by increasing the current) in proportion to the change in weight.

The accuracy of the results depends very much on the use of small samples, which have been well ground to a uniform particle size. Apart from being used to study the decomposition of pure materials, it may also be employed to determine the percentage composition of binary mixtures of known substances, because the individual weight loss at each step on the curve can usually be related to one or other of the two components.

Differential analysis
There are two other forms of thermal analysis. In differential thermal analysis (DTA), the material under examination is compared with a corresponding amount of reference material (usually alumina) while the two are being heated with a steady rise in temperature. As the heating progresses, any change in the sample results in the release or absorption of energy so that a temperature difference occurs between it and the reference. This difference, which DTA measures, gives rise to a graph consisting of a series of peaks (the exotherms) and troughs (the endotherms). The former correspond mainly to chemical changes in compounds, whereas the latter indicate physical changes in the crystalline structure or fusion of molecules. The widths of the peaks and troughs are also of importance in indicating the changes in the sample.

By contrast, differential scanning calorimetry (DSC) measures the difference in energy required to keep both the sample and reference substances at the same temperature. This means that when, for example, an endotherm would occur on DTA, in DSC that heat difference is made up by the

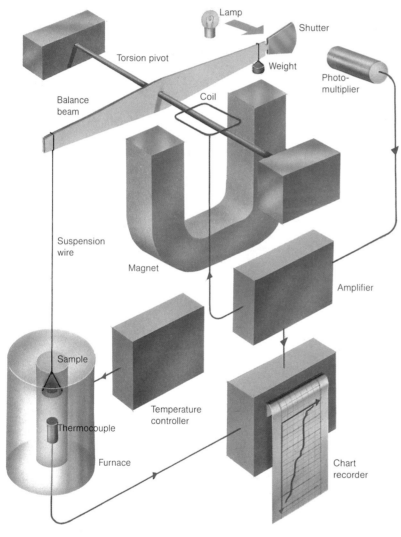

A thermal balance records the loss in weight of a substance as its temperature is gradually increased. A counterbalanced sample of the substance is suspended in a furnace and heated; a thermocouple detects the rise in temperature. As the sample loses weight, the other end of the balance beam falls, moving a shut-ter and allowing light to fall onto a photomultiplier. The signal from the photomultiplier is amplified and passed through a coil fixed to the torsion pivot of the balance beam. The coil is also situated in a magnetic field, and so the current flowing in it causes it to rotate slightly (like the armature of an electric motor), thus twisting the suspension, restoring the balance and making the shutter interrupt the light beam. The current in the coil is a measure of the weight lost by the sample, plotted by a chart recorder as a thermal gravimetric graph (see example on the opposite page).

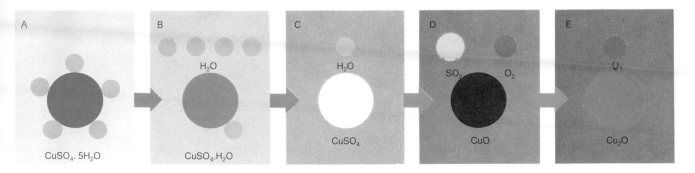

A

$CuSO_4 . 5H_2O$

B

H_2O

$CuSO_4.H_2O$

C

H_2O

$CuSO_4$

D

SO_2 O_2

CuO

E

O_2

Cu_2O

The thermal decomposition of copper sulfate is shown diagrammatically above. At room temperature, this salt exists as the pentahydrate, with five molecules of water of crystallization (A). On heating, it loses water in two stages to give first the monohydrate (B) and then the anhydrous salt (C). On continued heating anhydrous copper sulfate decomposes to copper (II) oxide (D) and then copper (I) oxide (E). Thermal analysis, which records weight loss with rising temperature, gives a graph (right) with "steps" corresponding to each of the reactions.

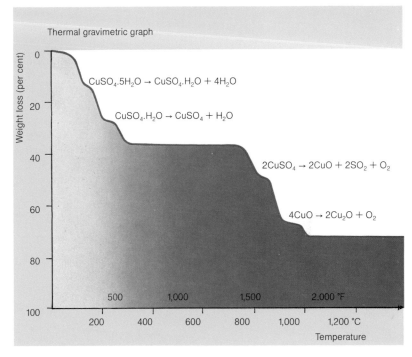

Thermal gravimetric graph

$CuSO_4.5H_2O \rightarrow CuSO_4.H_2O + 4H_2O$

$CuSO_4.H_2O \rightarrow CuSO_4 + H_2O$

$2CuSO_4 \rightarrow 2CuO + 2SO_2 + O_2$

$4CuO \rightarrow 2Cu_2O + O_2$

Weight loss (per cent)

Temperature

instrument adding more heat to the sample. In this case the instrument measures the amount of extra energy that has to be added to either the sample or reference container in order to maintain both at the same temperature. In all cases the shapes of the peaks obtained are affected by the rate of heating, the sample size and even the shape of the sample containers.

The equipment for DTA and DSC consists of an electrically-heated solid metal block which contains two identical recesses for the sample and reference capsules. The temperatures of the two substances are measured, as the block is heated, by two thermocouples. The whole system is operated according to a preset temperature program, often controlled by a microcomputer. Applications of thermal methods of analysis are expanding very rapidly. They have been used to determine deterioration in high alumina cement in buildings as well as for establishing the best curing temperatures for epoxy resins and the fusion temperatures between mixtures of inorganic compounds.

Thermoluminescence dating

A totally distinct form of thermal study is thermoluminescence dating, which is used to try to establish the age of archeological specimens. The method is based upon the idea that some materials, such as rocks and minerals, emit visible light when heated due to electron displacements taking place in the crystal lattice. These are believed to relate to the amount of exposure to natural radioactivity and are built up over an extended period of time so that the greater the amount of light released, the older are the rocks, whose age can be estimated from a plot of the light intensity against the temperature. The same principle is now employed in personnel badges used for monitoring exposure to radioactivity in laboratories.

A modern instrument for carrying out differential thermal analysis (below) has a computerized temperature controller and records the release or absorption of heat energy by a sample as it undergoes chemical changes or changes in its physical makeup.

Chromatography

Although chromatographic analyses have been known for nearly a century, it is only since the 1950s that chromatographic techniques have become widely used as general procedures. The basic principle underlying all such methods is that a substance in either liquid or gaseous solution (known as the mobile phase) passing down a packed column or over the surface of a solid material is slowed down in its progress due to its interaction with the stationary phase.

In effect the stationary phase acts as a molecular obstacle course, the molecules of one chemical species moving faster or slower than those of another depending upon their chemical nature and molecular sizes. What actually happens with the materials in solution is that there is competition for them between the mobile and stationary phases. The greater the affinity with the stationary phase, the slower the compound in solution moves; conversely, the greater the affinity for the mobile phase, the faster the compound moves. It is because of this different rate of movement that chemicals can be separated from each other and can be identified by the rates at which they move under specified conditions. The great value of chromatographic methods lies in the fact that the three analytical objectives of separation, identification and quantification can frequently be achieved simultaneously.

Types of chromatagraphy

The main forms of chromatography are classified into the two groups of adsorption chromatography and partition chromatography. In the former the stationary phase is simply a solid material of a uniform fine particle size upon which components in solution are progressively adsorbed and desorbed as they pass over the surface. But in partition chromatography the stationary phase is a viscous liquid of high boiling point which has been deposited on the surface of chemically inactive particles which serve simply as a support material. In this case separation of substances in the mobile solution depends upon the way in which they partition between the mobile and stationary phases.

Early chromatographic separations were applied mainly to colored compounds and their separated zones could be readily observed once the chromatograms had been run. However, many separations are now carried out on colorless substances. These are detected by passing the solution through special detector cells, which measure changes in properties such as conductivity, refractive index or ultraviolet absorption as the solutions pass through.

Column chromatography

The earliest form of chromatography, which is still regularly used, is that of adsorption chromatography in columns. It is easy to set up and can give good results both on a small and a large scale. Typically the procedure involves packing a glass tube 1 inch (25 millimeters) in diameter and 20 inches (50 centimeters) long with an adsorbent, such as powdered alumina. The mixture to be separated is added as a small concentrated volume at the top of the column, followed by the continuous addition of an appropriate solvent. The separated components of the mixture can either be run out (eluted) from the bottom of the column if sufficient solvent is added, or they can be obtained by removing the packed material from the tube and extracting the separated compounds from their zones. This form of chromatography is useful for large-scale operation in which sizable quantities of material are to be separated.

Plate, or thin-layer, chromatography (TLC)

Adsorption chromatography can also be carried out on a small scale by using an adsorbent in the form of a thin layer spread on a glass or aluminum foil sheet. A small volume of the sample to be separated is placed near one edge of the sheet, and that edge placed in a shallow pool of solvent at the bottom of a glass tank. As the solvent moves up the plate by capillary and molecular attraction, it carries the components in the mixture with it and these are separated according to their affinities for the alumina or silica gel sur-

In column chromatography, a mixed solution is poured into the top of the column, and a solvent trickled through to separate the mixture into bands. The various fractions can then be washed through one at a time using a series of other solvents.

Solvent

Moving solvent phase

Bands of separated components

Solid absorbed phase

1 2 3 4 5

← Fractions

face. The run can be stopped at any point and the movement of any component expressed in numerical terms as a ratio (the R_F value) with respect to the movement of the solvent front. Identification of compounds can be made by measuring their R_F values when separated by different solvent systems and by comparing them with reference substances.

Larger quantities of mixtures may be separated by using fairly thick layers of the adsorbents. The separated materials are obtained in a pure state by scraping off the appropriate part of the adsorbent surface and extracting the substance with a solvent such as ethanol.

Paper chromatography

This is now less commonly used than thin-layer chromatography, although it was widely employed before the general introduction of TLC. Because paper contains a high proportion of water, the separations carried out on it are considered to be mainly due to partition between the organic solvent mobile phase and the stationary water held in the fibers of the paper. One advantage of paper chromatography over TLC is that it can be run in either a descending or ascending manner. Although the latter is easier to carry out, the former is preferable because it is faster, due to the effect of gravity on the solvent flow. As with TLC, the sample is placed at a point a short distance from the edge of the paper, which is then dipped in a tray of solvent in a covered glass tank. As the solvent travels along the paper it carries the substances in the sample with it, and these are separated due to their different distributions between the mobile (solvent) and stationary (water) phases.

Gas chromatography

The introduction of partition gas chromatography by Tony James and Archer Martin in 1953 revolutionized chemical analysis because it introduced a process suitable for very wide application. In this form of chromatography the sample mixture (as a liquid or vapor) is injected by a syringe into a continuously flowing gas stream (nitrogen, argon or helium), which is passing along a narrow column. The column is packed with an inert solid powder, which acts as a support for a stationary nonvolatile liquid (a silicone oil or polyethylene glycol). The substances in the sample distribute themselves between the mobile and stationary phases according to their physical

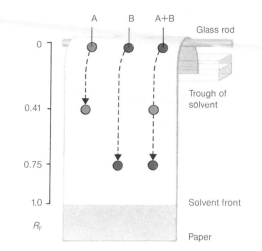

In descending paper chromatography, solvent passes down a sheet of filter paper which has been "spotted" at the top with the mixture to be separated. Each component moves a different distance, described by its R_F value (the ratio of the distance traveled by the component to the distance traveled by the solvent front).

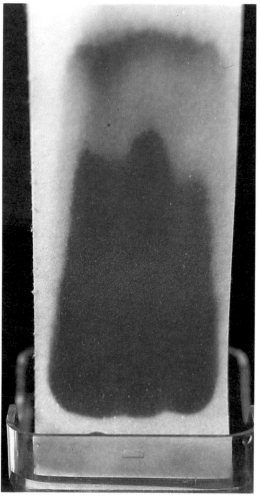

The simple ascending paper chromatogram (left) shows the results of making an elongated blot of black ink near the foot of a sheet of filter paper and placing the lower edge of the paper in a trough of solvent. On its way upward, the solvent has separated the ink into its various components.

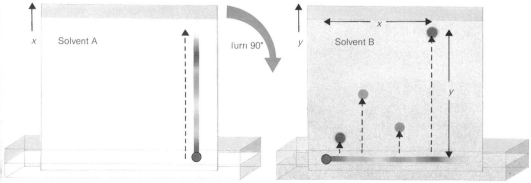

In two-phase chromatography (left), a mixture is "spotted" near one corner. Solvent A separates it into regions along a straight line (in the x direction). The paper or plate is turned through 90° in a second solvent B, causing each component to travel a different distance in the y direction.

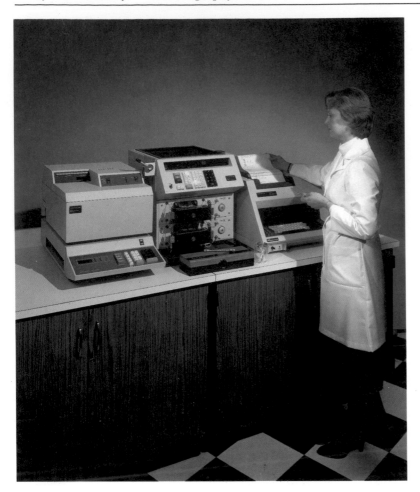

Liquid chromatography can be automated under computer control using the apparatus illustrated above, which prints out the results on a paper chart.

and chemical properties, which separate them, and they are successively carried out by the gas stream at the end of the column. The presence of the individual substances is measured by special electronic detectors at the end of the column, and the whole sequence recorded as a series of peaks on a chart. Individual chemicals can be identified by the relative times they take to pass through the column (the retention time). Also the areas of the peaks can be used to measure quantitatively the amount of material detected. By this means very complex mixtures of oils, solvents, drugs and pollutants can be readily separated.

Gas chromatography and mass spectrometry
One very useful application of gas chromatography has been in linking it with a mass spectrometer. The process is known as interfacing, and necessitates removing most of the carrier gas from the separated samples before they can pass to the mass spectrometer. The interface is designed so that whenever the chromatography detector shows that a separated substance is coming off the column, part of the effluent is diverted to the mass spectrometer, where it produces a mass spectrum. The substance can be identified by comparing its spectrum with those previously determined (in modern instruments this can be done almost instantaneously by computer). As a result, the combined technique has become especially useful for the qualitative analysis of complex organic mixtures.

High performance liquid chromatography
Probably the most rapidly expanding area in chromatography, HPLC incorporates all standard forms of chromatographic analysis and can be applied to most chemical substances. In practice it is conventional adsorption, partition, exclusion or ion-exchange chromatography carried out on microcolumns with the solvents pumped through under high pressures of up to 600 atmospheres. The great advantages of HPLC are that only very small samples are required, the detectors used (based upon fluorescence, ultraviolet absorption or refractive index) are highly sensitive, and a very high degree of separation can be achieved between closely related substances.

Preparative chromatography
Traditional methods of separative chromatography normally deal with very small volumes of samples, so that only minute amounts of pure materials are obtained. Demands for larger quantities of very pure compounds for use as standards and the determination of correct chemical constants have, however, necessitated the development of what are called preparative chromatographic methods. These fall into the two categories of large-scale chromatographic separations or multiple small-scale operations.

Typical of the first type are the thick-layer chromatograms in which the traditional adsorbent layer is made up to 0.4 inch (1 centimeter) thick, using sample leads in excess of $\frac{1}{28}$ ounce (1 gram). Similarly, gravity-flow adsorption and partition chromatographic columns can be scaled up to 4 inches (10 centimeters) diameter and $6\frac{1}{2}$ feet (2 meters) long, with comparatively large sample loadings. The limitations in all these methods are that the quality of the separation between the various zones of the column suffers, and the maximum sample loading and column length depend upon how pure the required extracted chemical has to be.

The second type of preparative chromatography uses gas chromatography in which many small sample volumes, usually in the order of

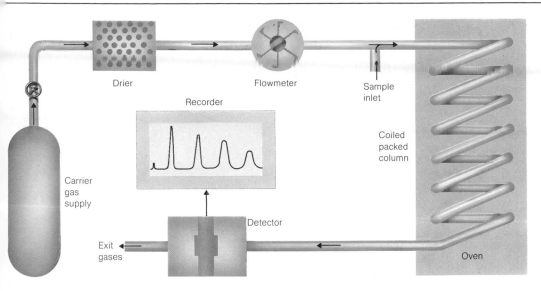

In a gas chromatograph, a carefully monitored flow of a dry, inert carrier gas (such as helium or argon) carries samples to be analyzed through a coiled column heated in an oven. The same gas carries the separated components to a detector, which usually responds to changes in thermal conductivity of the gas, recorded on a graph. An example of a gas chromatogram is shown below for a mixture of the four isomers of butanol.

microliters at a time, are injected and the even smaller quantities of pure, separated chemicals collected sequentially in a cold trap. The system is best when it is automated, with a regular injection of sample followed by a sequence of traps at the end of the column as individual compounds are eluted. Although rather time-consuming, it is a useful method for obtaining small quantities of very high-purity chemicals.

Other methods of separation

Although not strictly chromatographic, similar separation methods used in analytical chemistry include ion exchange and electrophoresis. Ion exchange uses a column packed with a special synthetic resin (a sulfonated styrene copolymer) or a natural substance such as zeolite, which "exchanges" its ions with those of a solution passed through the column. The most familiar application of the technique is in water softeners, which take up calcium ions from hard water and exchange them for sodium ions.

Electrophoresis is more like chromatography. It involves the movement of chemicals in gel columns or on paper or cellulose acetate strips. But in electrophoresis, the movement of sub-

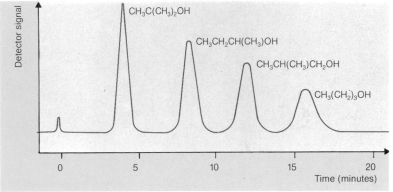

stances relative to the stationary phase is brought about by applying an electric field along the length of the strip or column, with the result that negatively-charged ions and polarized molecules move towards the anode, and positively-charged ones move towards the cathode. The extent of the movement for different substances under the influences of the applied potential depends upon the charges, sizes and shapes of the substances. This method of separation is of particular value in the study of biochemicals such as proteins, peptides and amino acids.

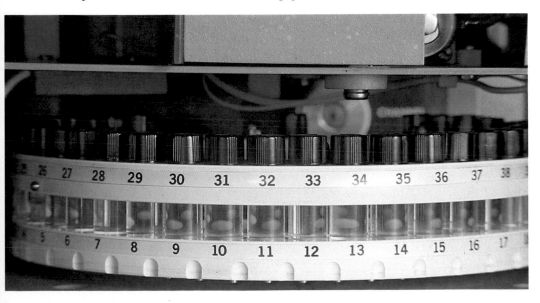

A key part of a gas chromatograph is a heated coil (far left), packed with a solid adsorbent coated with a nonvolatile liquid. The turntable (left) enables many samples to be injected sequentially into the carrier gas flowing into the chromatograph.

Automated analysis

Increasing demand for chemical analysis in hospitals, for environmental monitoring and for quality control has led to the modification of many analytical methods so that the analysis can be carried out automatically for dealing with large numbers of similar samples. Results obtained by automatic means are often stored in computers, and printed out on tables and record sheets.

The equipment available can be classified into three main groups. These are automatic instruments in which one or more steps are controlled by the instrument, automatic analyzers for multiple-sample treatment, and on-line continuous analysis.

Automatic devices
The first instruments for automatic analysis included such systems as automatic titrimeters,

which switch off when the end point of a titration has been reached, or reverse and automatically refill the syringe burette ready for the next titration. With this type of system the control of the process is still in the hands of the chemist, but part of the decision making is carried out by the instrument.

Many spectrophotometers now also incorporate devices which, at the end of scanning one spectrum, automatically wind on the chart paper, reverse the wavelength scale, and change the energy source (if necessary) ready for running the next spectrum. They all operate by means of an internal circuit which, at the end of first analysis, feeds back an electric signal to a control panel which activates the reversing and reset procedures. The whole sequence is carried out without the aid of the operator.

Automatic analyzers
The handling of large numbers of similar samples has been transformed by the introduction of automatic discrete sample analyzers. In this type of instrument the samples are loaded onto a turntable and a measured portion of each is withdrawn — one at a time at regular intervals — by means of a sampling probe. For spectrophotometric analyses each sample is pumped through a continuous flow system and separated from other samples by air bubbles. As the samples are pumped along the tube they are mixed with appropriate reagents supplied by tubes from storage bottles. After allowing sufficient time for any reaction to be completed, any color produced in the sample-reagent mixture is measured as the solution traverses a flow-through cell. The depth of color in the solution is automatically measured, recorded and listed under the appropriate sample number.

Automatic sampling syringes are also used for multiple analyses by gas chromatography. The sample turntable rotates to provide the next sample for analysis only when the first has been com-

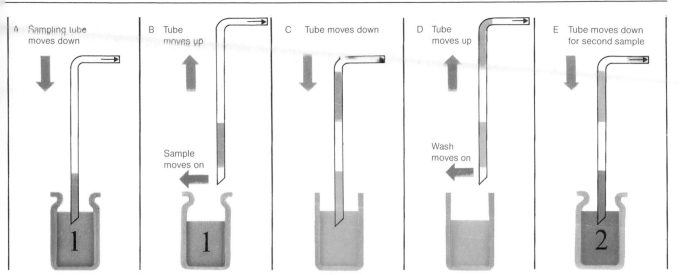

A Sampling tube moves down

B Tube moves up — Sample moves on

C Tube moves down

D Tube moves up — Wash moves on

E Tube moves down for second sample

pletely eluted from the column, because it is not normally possible to have two samples passing through the chromatographic column simultaneously. One of the main areas to benefit from the development of automatic analyzers is clinical analysis in medicine, in which the rapid and accurate analysis of body fluids for diagnosis can be critical for the patient.

Continuous flow analyzers

On-line monitoring, control and analysis instruments vary widely because most of them are designed for specific purposes. The simplest online instruments may be merely pH electrodes (for measuring acidity) or thermocouples (for monitoring temperature) dipping into a liquid in a pipeline or reaction mixture and arranged to give a signal to switch off the flow if the acidity or temperature varies between fixed limits. A similar arrangement may be used to give a continuous signal, which can be fed to a pen recorder to produce a permanent chart record of any variations that have occurred.

Continuous monitoring of this type is used extensively in pollution control and environmental studies. For example, an oxygen-selective ion electrode can be used to monitor continuously the dissolved oxygen at a point in a river or lake, a nondispersive infrared gas analyzer can be used to study carbon monoxide in the atmosphere, and other probe methods can be applied for a variety of heavy metal pollutants in water.

The use of on-line analyzers to control chemical and manufacturing processes is often, however, very complex. This is partly because there is a need to ensure the reliability of the analytical results even when the monitoring is carried out in adverse environments of high acidity, corrosive gases and elevated temperatures. Where possible, such analytical systems are based on continuous monitoring, with remote sensing devices that detect changes in magnetic field strength, turbidity of solution, color of reagents, or refractive index (or some other physical property). In all cases the signals obtained may be fed into computer-operated controls which adjust the flow of reagents,

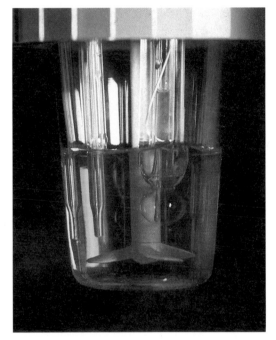

automatically raise or lower the temperature, terminate the reaction, or discharge the products.

Computerized control systems

The development of large-memory microcomputers has enabled even simple laboratory equipment to be linked with a computer to collect results from analyses as the bases for data storage, comparison and retrieval. As a result, computers are used at all levels of activity in chemical analysis. In spectrometry, they compare spectra from unknown substances with a library of spectra as a basis for identification. In chromatography, identification is achieved by the comparison of peaks with known retention times, and complete chromatographs can be recalled from the computer memory and presented on a display screen.

In this way, computers are effectively transforming the analytical laboratory. On the basis of the information fed into them by automatic instruments, they can indicate poor quality materials, suggest when a process is faulty, and even control the whole of a production sequence.

In one type of automatic sampler, a tube sucks up sample 1 (A, above left), then moves up and sucks in some air (B), moves down and sucks up wash liquid (C), moves up again to suck in air (D) and finally moves down to repeat the sequence with sample 2 (E). The samples are thus separated by bubbles of air and wash liquid.

A set of electrodes (left) can monitor the pH (acidity or alkalinity) of samples or reaction mixtures and feed the information to a computer.

Uses of analysis

The increasing demands for quality control in factories, higher standards for the cleanliness of air and water, and tighter requirements on drugs and food additives have all led to a greater need for more and better methods of chemical analysis.

Analysis of food and drugs

Chemical analysis has a major role in helping to maintain a high standard of reliability for the substances we eat, drink and use to maintain our health. Many chemicals are added to foods as preservatives, colorants, flavors, emulsifiers and humectants, and for this reason it is essential that regular tests are carried out to ensure that only permitted food additives — in the correct quantities — are incorporated. The most sensitive analytical procedures such as gas chromatography and ultraviolet absorption spectrometry are employed for this purpose, and they are also able to detect the presence of minute amounts of chemical impurities or adulterants.

Food and drug analysis is a specialized field, because the chemist often has to deal with complex mixtures of natural products — from fruits, cereals or animals — as well as with synthetic compounds. Often this entails resorting to complex extraction procedures before full identification by sophisticated procedures such as spectrometry. And without such regular analytic checks, the quality of food might be much poorer and harmful additives more likely to be used.

One area of considerable importance in food analysis is measuring the alcohol content in beers, wines and spirits. Many countries charge a tax or duty on such drinks based upon the alcohol content, so that accurate methods of measurement — often employing specific gravity — are important for levying the correct amount of duty.

Clinical analysis

Diseases and other medical disorders lead to changes in the concentration and production of various chemicals in the human body. In the diagnosis of medical conditions, clinical analysis plays an increasingly important part. For example, simple chemical tests for glucose in the blood or urine can reveal whether or not a person is diabetic. Flame emission spectrometry is used extensively for measuring sodium, potassium and calcium concentrations, and other spectrometric methods are employed for measuring urea. An imbalance of any substance can indicate bodily disorders. The demand for chemical analyses for medical purposes is now so great that automatic methods of analysis are used whenever possible.

Forensic analysis

An area of growing importance for chemical analysis is in the detection of crime and the conviction of criminals. For this purpose almost every analytical tool is called into play. It is, for instance, possible to determine whether or not a fire has been started deliberately, by analyzing the ashes and debris for the presence of kerosene, petrol or solvents. Gas chromatography is used for this purpose and can detect not only the main materials used to start the fire but also the identity of its individual components. Similarly the conviction of illegal drug sellers depends on correct identification of what are often apparently only white powders. To differentiate correctly between innocent sugar, talcum powder and flour, and the more lethal morphine, heroin and cocaine it is necessary to use chromatographic methods in conjunction with infrared and mass spectrometry.

There is no end to the variety of analyses forensic scientists have to carry out, including the identification of glass fragments, the chemical nature of paint flakes and the composition of inks. One of the major tasks in many countries is also the regular chromatographic analysis of blood samples to determine the alcohol (ethanol) content for motorists involved in drink-drive incidents. In cases such as this the analyst may have to be prepared to stand up in court to explain his results.

Quality control

One of the major uses of chemical analysis is, not surprisingly, in the chemical industry itself and in manufacturing for quality control. The demand for purer chemicals and materials has in turn led to great improvements in analytical methods. As chemicals are manufactured, their rate of production is often checked automatically by on-line analytical equipment fitted into the flow pipes. The final material also undergoes rigorous tests to ensure that it is not only the correct material but also that it satisfies the specification for the quality required. Even minute amounts of impurities can cause it to be rejected as unsuitable.

Such quality control is of immense importance

Pollution control is one of the major uses of chemical analysis. Here a scientist collects samples of river water from a feeder stream, which has a suspicious scum that may be caused by industrial or agricultural pollution.

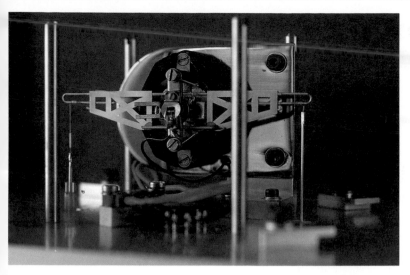

with chemicals that are used for human consumption, such as pharmaceuticals and cosmetics. Even household materials such as polishes, bleaches, detergents and paints all require proper analytical checks to ensure that the correct amounts of chemicals have been mixed. In a similar way, the metals used to make such products as cars and aircraft must meet tight quality specifications, because the presence of any impurities can cause corrosion to occur very rapidly. In this area spectroscopic and surface analysis methods are of prime importance. Building materials — bricks, cement and plaster — and adhesives all come into the realm of substances that deteriorate rapidly if they are incorrectly formulated, and careful quality control is necessary to maintain high standards.

Chemical analysis is the unnoticed essential step in maintaining the quality of most of the products that we use. Without analytical methods, it would be impossible to have confidence in the objects we all buy.

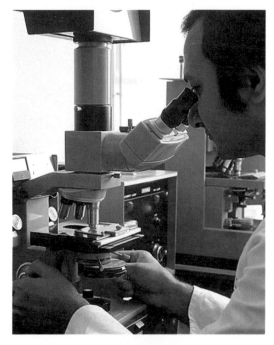

Accurate weighing remains the basis of most quantitative techniques, if only in the preparation of samples for subsequent automatic analysis. Delicate mechanical precision is combined with modern electronics in the mechanism of a modern balance (top left), whereas the simpler machine (above) displays a direct digital reading.

In forensic analysis, microscopic examination (left) often precedes chemical treatment of samples from the scene of a crime.

Despite the revolution in chemical analysis being brought about by the use of sophisticated instruments and computers, there will probably always remain a need for individual laboratory techniques in connection with most industries, for there are hardly any industrial processes that do not require the services of chemistry at some stage or other.

GLOSSARY

In the following glossary, small capital letters (e.g. RADIATION) indicate terms that have their own entries in the glossary.

A

absorption A process in which a substance takes up another substance, such as blotting paper (a solid) absorbing water (a liquid), or in which a substance takes up RADIATION of characteristic WAVELENGTHS. The latter process gives rise to absorption spectra, of importance in chemical ANALYSIS. *See also* ADSORPTION; SPECTROSCOPY.

acid A substance from which a hydrogen ion can be released in aqueous SOLUTION. The strength of an acid depends on the ease with which the hydrogen ion dissociates from the rest of the molecule and is measured by its PH, which is always less than 7. *See also* PHASE; CARBOXYLIC ACID; HYDROGEN CONCENTRATION.

actinide One of a group of ELEMENTS which starts with actinium (element 89) and concludes with neptunium (element 103). Like the LANTHANIDES, the actinides form a separate group within the PERIODIC TABLE, in which an inner SHELL of ELECTRONS is being filled as the ATOMIC NUMBER increases.

activation energy The energy needed in order to form a TRANSITION STATE, the unstable intermediate which occurs during a chemical REACTION. This intermediate is at a higher energy level than either the reactant(s) or product(s) and its formation consequently requires the surmounting of an energy barrier.

addition reaction A chemical REACTION in which the product is formed by combining two separate MOLECULES. It most frequently refers to the breaking of a multiple BOND between two carbon ATOMS and the subsequent formation of two new single bonds to other atoms.

ADP *See* ATP.

adsorption A process in which a substance adheres to the surface of another substance. Adsorption is important in some types of CATALYSIS, notably where gases adsorb on METAL surfaces and REACTION is facilitated by a consequent lowering of ACTIVATION ENERGY. *See also* ABSORPTION.

alcohol An ORGANIC chemical substance which contains a HYDROXYL FUNCTIONAL GROUP. Sometimes taken to refer specifically to ethanol (ethyl alcohol), the intoxicating constituent of alcoholic beverages.

aldehyde An ORGANIC chemical substance in which one of the terminal carbon ATOMS has an oxygen atom attached to it by a double BOND and a hydrogen atom attached to it by a single bond. *See also* KETONE.

alicyclic A type of ORGANIC chemical substance in which the carbon ATOMS are linked together end to end to form a ring structure which either contains no double bonds or insufficient to achieve AROMATICITY. *See also* ALIPHATIC.

aliphatic A type of ORGANIC chemical substance consisting primarily of a chain of carbon atoms. *See also* ALICYCLIC; AROMATICITY.

alkali A basic substance, particularly a hydroxide salt. *See also* BASE.

alkaloid Any one of a large family of ORGANIC compounds found in plants and having a basic nitrogen ATOM incorporated into a HETEROCYCLIC structure. Most of them are physiologically active.

alkane A saturated ALIPHATIC HYDROCARBON in which all the carbon and hydrogen atoms are joined to each other by single bonds; formerly known also as a PARAFFIN.

alkene An ALIPHATIC HYDROCARBON that contains at least one carbon-carbon double bond (for example, ethene or ethylene); formerly known as an olefin.

alkyl group An ORGANIC chemical group derived from a HYDROCARBON and consisting solely of carbon and hydrogen ATOMS. The group is able to form a single bond between one of its carbon atoms and another atom, or group of atoms, to form more complex compounds.

alkyne An ALIPHATIC HYDROCARBON that contains at least one carbon-carbon triple bond.

allotropy The occurrence of a chemical ELEMENT in more than one physical form as a result of the formation of HOMOGENEOUS MOLECULES containing different numbers and/or arrangements of ATOMS. Carbon, phosphorus, sulfur and many METALS occur in different allotropic forms.

alloy A mixture of two or more METALS, combined in order to produce a material with more useful properties than any of the components individually. Thus, elements such as niobium, which cannot be formed easily into complex shapes but which are hard, are alloyed with iron, which is malleable, to produce hard steels.

alpha particle The NUCLEUS of a helium ATOM, made up of two PROTONS and two NEUTRONS. This particle is emitted during the breakdown of some RADIOACTIVE ELEMENTS. *See also* IONIZING RADIATION.

amine A COMPOUND in which one or more of the hydrogen atoms in ammonia have been replaced by ORGANIC chemical FUNCTIONAL GROUPS, such as ALKYL GROUPS.

amino acid An ORGANIC chemical substance in which one carbon ATOM has attached to it both an amino group ($-NH_2$) and a CARBOXYLIC ACID group. Amino acids are the MONOMERS from which PROTEINS are built.

analysis The determination of the composition and structure of chemical COMPOUNDS and of the composition of mixtures of different compounds.

anion An ION which has a greater number of ELECTRONS associated with it than there are PROTONS in its atomic NUCLEUS (or nuclei) and which consequently carries one or more units of negative electric charge.

anode A positive ELECTRODE, that is, one towards which ANIONS move during ELECTROLYSIS.

antibiotic An ORGANIC substance, usually with a complex chemical structure, produced by a MICROORGANISM which has a harmful effect on some other microorganisms.

antibody A complex PROTEIN produced by an animal cell in response to attack by a foreign body such as a MICROORGANISM. The antibody reacts with an antigen, which may be the protein coat of a VIRUS for example, and helps to destroy the invading organism. Each antigen stimulates production of a particular antibody which reacts only with that antigen.

argonon Any one of the ELEMENTS, helium, neon, argon, krypton, xenon and radon, the family which occupies the column on the far right of the PERIODIC TABLE. They are characterized by having a full complement of *s* and *p* electrons in their outermost SHELL. Because they are very unreactive they used to be known as the inert or noble gases; they are often still referred to as the rare gases.

aromaticity The phenomenon in which some of the ELECTRONS linking carbon ATOMS together into ring structures are shared by all the atoms in the ring, occupying DELOCALIZED ORBITALS. Delocalization affects the chemical behavior of the COMPOUNDS, so that they differ in reactivity from related ALIPHATIC or ALICYCLIC compounds. The "parent" compound is benzene. It and some of its close relatives were first isolated in large quantities from coal tar and their characteristic odor led to the name "aromatic."

aryl group An ORGANIC chemical group derived from an AROMATIC HYDROCARBON and, consequently, consisting solely of carbon and hydrogen atoms. The group is able to form a single BOND between one of its carbon ATOMS and another atom, or group of atoms, to form more complex compounds, such as aryl halides.

asymmetry In chemistry, asymmetry usually refers to a carbon ATOM in an ORGANIC compound which is linked by single chemical BONDS to each of four different atoms or groups of atoms. Compounds which contain an asymmetric carbon atom can occur as different STEREOISOMERS and are OPTICALLY ACTIVE.

atom The smallest amount of any ELEMENT which can exist, while still having the properties characteristic of that element.

atomic number The number of protons in the nucleus of an ATOM of an ELEMENT. As each element has a different number of nuclear protons, the atomic number is a fundamental characteristic of an element.

ATP Adenosine triphosphate, an ORGANIC compound used by living organisms as a means of transferring energy between other compounds. The basic reaction of ATP is the transfer of one of its phosphate groups to another molecule, accompanied by the formation of adenosine diphosphate (ADP). The addition of a phosphate group to the other molecule facilitates its part in further reactions. The ADP is rephosphorylated to ATP at a different stage of the metabolic process by other compounds, which lose energy as a result.

B

base A substance which will react with an ACID to form a SALT. In addition to ALKALIS, which are frequently metal hydroxides, bases include a number of ORGANIC nitrogen compounds. An aqueous solution of a base always has a PH greater than 7.

beta particle An energetic ELECTRON, emitted from a NUCLEUS when a NEUTRON changes into a PROTON during RADIOACTIVE breakdown of an ATOM.

biochemistry The branch of chemistry that is concerned with the chemical REACTIONS and substances which occur in living organisms.

bond The link between two ATOMS in a MOLECULE, which consists of a sharing of ELECTRONS in a molecular ORBITAL.

C

calorie A measure of heat energy, now largely obsolete. Formally, it is the amount of heat required to raise the temperature of 1 gram of water by 1° centigrade. It has now been replaced in scientific usage by the JOULE (1 joule = 4.18 calories). The calorie is still used as a measure of the energy availability of foods; this calorie, however, is really a kilocalorie (that is, 1,000 calories).

carbohydrate An ORGANIC chemical compound made up solely of carbon, hydrogen and oxygen atoms. Usually, the ratio of hydrogen to oxygen atoms is 2:1, as in water (H_2O). Carbohydrates include many compounds important in living processes, such as SUGARS, STARCH and CELLULOSE.

carbonyl group An ORGANIC chemical group in which an oxygen ATOM is doubly-bonded to a carbon atom which in turn is also bonded to other carbon or hydrogen atoms. KETONES and ALDEHYDES are the main classes of COMPOUND containing a carbonyl group.

carboxyl group An ORGANIC chemical group in which a carbon ATOM is bonded to two oxygen atoms by a total of three chemical BONDS. CARBOXYLIC ACIDS are the main class of COMPOUNDS containing the carboxyl group.

carboxylic acid An ORGANIC chemical COMPOUND in which a terminal carbon ATOM has an oxygen atom and a HYDROXYL GROUP linked to it and which is ACIDIC because the hydrogen atom of the hydroxyl group can dissociate from the oxygen as a positive ION. *See also* CARBOXYL GROUP.

catalysis The process by which a nonreacting substance (the catalyst) lowers the ACTIVATION ENERGY of a REACTION involving one or more other substances. This means that the reaction can proceed faster, under milder conditions, than it would in the absence of the catalyst.

cathode A negative ELECTRODE, that is, one towards which CATIONS move during ELECTROLYSIS.

cation An ION which has a smaller number of ELECTRONS associated with it than there are PROTONS in its atomic NUCLEUS (or nuclei) and which consequently carries one or more units of positive electric charge.

cellulose A CARBOHYDRATE POLYMER that occurs widely in plants as a structural material. Cotton and paper are both made from cellulose fibers.

chain reaction A chemical REACTION in which the energy or substances produced encourage more of the starting materials to react, so that the reaction becomes self-sustaining. If the effect is very great, the reaction may become explosively fast. Chain reactions can also occur at the subatomic level, as in the nuclear fission process in an atomic bomb explosion.

chelation The property of some molecules physically to hold onto METAL IONS, usually by the attraction between the ion's positive charge(s) and the LONE PAIR electrons of nitrogen or oxygen atoms. Chelating agents can be useful in removing unwanted metal ions from solutions and for identifying particular ions in analytical chemistry.

chemiluminescence Light generated by a chemical REACTION, generally without any accompanying heat. *See also* FLUORESCENCE; PHOSPHORESCENCE.

chirality The property of "handedness" in molecules. Chiral atoms are those which are ASYMMETRIC and consequently give rise to STEREOISOMERS. *See also* OPTICAL ACTIVITY.

chromatography A technique for separating similar chemical compounds from one another, by using differences in the strength of their ADSORPTION on an inert material. It is widely used in analytical chemistry and also in the separated purification of some high-value compounds.

chromophore A part of a MOLECULE which can interact with light, absorbing some WAVELENGTHS selectively and thus making the molecule colored.

codon A group of three organic bases in a stretch of DNA which codes for a specific AMINO ACID.

coenzyme A small nonprotein molecule which combines with the PROTEIN of an ENZYME and which is essential to its function. Not all enzymes require coenzymes. Some VITAMINS are needed in the diet because they act as coenzymes.

colloid A dispersion of one type of MOLECULE in another which has some, but not all the properties of a SOLUTION. Large molecules, such as PROTEINS, form colloidal dispersions. In some cases, the dispersion appears to be solid, as in the case of a jelly — a colloidal suspension of the protein gelatin in water. EMULSIONS are colloidal dispersions of two immiscible liquids. Colloidal dispersions of small molecules can be formed from materials which are normally insoluble; colloidal stability is maintained by careful balancing of IONIC charges.

compound A substance made up of heterogeneous MOLECULES, that is, molecules in which ATOMS of at least two different ELEMENTS are bonded together.

concentration The amount of a particular substance present in a defined volume, used most frequently to refer to the amount of a solid or liquid substance dissolved in another liquid. Concentration is also used to mean the act of making a SOLUTION more concentrated.

condensation reaction A chemical REACTION in which two MOLECULES link together, usually with the expulsion of a small molecule, such as water or ammonia.

conductivity A measure of the ability of a material to pass an electric current. *See also* SUPERCONDUCTIVITY.

conformation The shape of a MOLECULE in space. In particular, it often refers to the shape adopted by six-membered carbon rings in complex ORGANIC compounds.

coordination compound A MOLECULE in which at least one BOND between two ATOMS is made up of a pair of ELECTRONS derived solely from one of them. Frequently, coordination compounds are made up from TRANSITION ELEMENTS bonded coordinately to several groups each containing an atom with a LONE PAIR of electrons.

corrosion The breakdown of METALLIC materials under natural conditions, such as the rusting of iron.

covalency The sharing by two ATOMS of pairs of ELECTRONS to form one or more chemical BONDS, with equal numbers of electrons originating from each atom.

cracking The breaking down and reforming of HYDROCARBONS from crude oil. By controlling the cracking conditions, the natural composition can be altered to produce larger amounts of particular hydrocarbons.

D

dalton The unit of ATOMIC and MOLECULAR mass, defined as one-twelfth of the mass of a neutral ATOM of the carbon ISOTOPE which contains six PROTONS and six NEUTRONS in its nucleus. It is named after the eighteenth-century English chemist, John Dalton.

delocalized orbital A molecular ORBITAL which encompasses more than two ATOMIC NUCLEI.

detergent A substance which has a cleansing action in solution, generally because it is a surface-active agent. *See also* SURFACTANT.

DNA Deoxyribonucleic acid, a POLYMERIC material found in living organisms. It carries the GENETIC CODE according to which organisms develop in an appropriate way (for example, so that a carrot cell will always behave like a carrot cell and not start developing into a cabbage, say) and transmits this information between generations. *See also* CODON; NUCLEIC ACID; NUCLEOTIDE; RNA.

E

electrode A material used to pass an electric current through a liquid or gas. A positive electrode is called an ANODE, a negative electrode a CATHODE. *See also* ELECTROLYSIS.

electrolysis The breakdown of a COMPOUND, usually in SOLUTION or as a liquid, when an electric current is passed through it. Under the influence of the current, the compound ionizes and the negative IONS (ANIONS) migrate to the positive ELECTRODE (ANODE), where their electrical charge is neutralized and they are either liberated or react with the surrounding medium. The positive ions (CATIONS) migrate to the negative electrode (CATHODE) where a similar charge NEUTRALIZATON takes place.

electromagnetic spectrum The band of WAVELENGTHS of electromagnetic radiation. It covers (from short to long wavelengths) GAMMA RAYS, X RAYS, ULTRAVIOLET RADIATION, visible light, INFRARED RADIATION, MICROWAVES and radio waves. Because the energy associated with radiation decreases with increasing wavelength, the different types of radiation interact differently with substances.

electron The smallest discrete subatomic particle in an atom. It has a negative electric charge and a mass nearly 2,000 times less than that of the PROTON and NEUTRON. Nevertheless, it is the constituent of ATOMS which contributes most to their chemical properties.

electronegativity A measure of the power with which an atomic NUCLEUS attracts ELECTRONS involved in a chemical BOND. The more electronegative an ELEMENT, the greater the IONIC character of the bonds it forms.

electroplating A process in which a thin layer of METAL is deposited on the surface of another metal by ELECTROLYSIS. The metal which is to be deposited is held in SOLUTION as IONS and the metal to be coated acts as a CATHODE.

electropositivity A measure of the ease with which an atomic NUCLEUS gives up ELECTRONS to form a CATION. The reverse of ELECTRONEGATIVITY.

element A substance composed of ATOMS whose NUCLEI all contain the same number of PROTONS. Elements are the basic atomic building blocks of matter; just over 100 of them are known.

emulsion An intimate mixture of two liquids which will not dissolve in each other, but which will not separate of their own accord. Such mixtures are a type of COLLOID.

endothermic reaction A chemical REACTION which has to absorb energy (not necessarily as heat) in order to take place. *See also* EXOTHERMIC REACTION.

enzyme A large ORGANIC MOLECULE, consisting mostly or entirely of PROTEIN, which acts as a CATALYST in a living system.

equilibrium The balance of a chemical REACTION. In any reacting system, provided that materials and energy are not added or taken away, there is always a fixed proportion of reactants and products which will be achieved if the system is kept undisturbed for long enough. This is known as its chemical equilibrium. Much of practical chemistry is aimed at disturbing the equilibrium of reactions to produce extra quantities of desired products, and of altering the rate at which equilibrium is reached by the use of CATALYSTS.

esterification reaction A REACTION in which an ACID, usually ORGANIC, and an ALCOHOL link together, with the expulsion of a molecule of water. Esterification is a type of CONDENSATION REACTION.

exothermic reaction A chemical REACTION which liberates energy, usually but not necessarily as heat. *See also* ENDOTHERMIC REACTION.

F

fermentation The microbiological process in which a sugar is converted to ethanol (ALCOHOL) by yeast cells. Nowadays, the term is often used to refer to the industrial production of any useful material (for example, an ANTIBIOTIC) by a MICROORGANISM.

fluorescence A phenomenon exhibited by various substances in which they absorb light of a particular WAVELENGTH and emit light of a different wavelength.

free radical An ATOM or group of atoms in which there is an unpaired ELECTRON, which makes the radical highly reactive. Free radicals are formed when a chemical BOND breaks in such a way that one electron remains with each molecular fragment. They are often important intermediates in reactions, particularly CHAIN REACTIONS.

functional group A group of ATOMS linked by chemical BONDS which confers a particular character on a MOLECULE and which may be able to participate in chemical reactions as if it were a single atom, for example, the HYDROXYL GROUP (—OH).

G

gamma ray Short WAVELENGTH ELECTROMAGNETIC radiation emitted during the RADIOACTIVE decay of some elements. *See also* IONIZING RADIATION.

gene A length of DNA which contains the code for the manufacture of a particular PROTEIN. In most organisms, genes are strung together into longer DNA units called chromosomes.

glycoprotein A complex ORGANIC substance in which a SUGAR is linked chemically to a PROTEIN.

H

half-life The time taken for half of any amount of a RADIOACTIVE ISOTOPE to decay. The half-life is characteristic for each isotope and, for different isotopes, varies between small fractions of a second and many thousands of years.

halide A chemical substance which contains one or more ATOMS of the halogen elements: fluorine, chlorine, bromine, iodine and astatine.

heavy metal Any of the metallic elements beyond sodium in the PERIODIC TABLE.

heterocycle An ORGANIC chemical COMPOUND containing one or more rings of ATOMS and having at least one atom other than carbon in one of the rings.

hexose A simple SUGAR containing six carbon ATOMS, for example, glucose. *See also* CARBOHYDRATE.

homogeneous Having a uniform makeup. It may refer to a MOLECULE in which all the ATOMS are the same or to a mixture, such as a SOLUTION, in which the constituent parts are thoroughly and evenly mixed.

hormone A physiologically active substance secreted by the endocrine glands in animals that produces an effect elsewhere in the body. A number of similarly active materials produced by plants are also loosely referred to as hormones.

hydrocarbon An ORGANIC chemical substance consisting solely of carbon and hydrogen ATOMS. Saturated hydrocarbons have all their carbon atoms joined by single BONDS and their remaining bonding capacity taken up by hydrogen atoms. Unsaturated hydrocarbons have one or more carbon-carbon multiple bonds. Hydrocarbons may be ALIPHATIC, ALICYCLIC or AROMATIC. They are obtained primarily from crude oil and are among the most important raw materials of the chemical industry.

hydrogen bond A weak BOND between a hydrogen ATOM in a MOLECULE and an unshared pair of ELECTRONS in another atom, such as oxygen. Although the hydrogen bond is only a fraction as strong as an ordinary COVALENT bond, it plays an essential role in determining the shape of many complex molecules of biological importance, such as DNA.

hydrogen ion concentration The concentration of hydrogen IONS in a SOLUTION, which dictates whether the solution is ACIDIC, BASIC or neutral. Many compounds dissociate either completely or partly in solution into electrically charged ions. Even pure water, which consists of MOLECULES of hydrogen oxide (H_2O), contains some dissociated hydrogen (H^+) and hydroxyl (OH^-) ions. As the amounts of each type of ion are equal, pure water is a neutral liquid. An excess of hydrogen ions makes a solution acidic, whereas a basic solution has an excess of hydroxyl ions. The hydrogen ion concentration of a solution is expressed as a mathematical function called the PH. A pH of less than 7 indicates an acid solution, one of greater than 7 an alkaline or basic solution.

hydrophilic Having an affinity for water. Hydrophilic substances dissolve readily in water. They are either IONIC compounds or ones which contain POLAR groups which interact with the highly polar water MOLECULE. *See also* HYDROPHOBIC; SURFACTANT.

hydrophobic Having an aversion to water. Hydrophobic substances, such as oil, do not dissolve in water because their MOLECULES are virtually nonpolar. *See also* HYDROPHILIC; SURFACTANT.

hydroxyl A FUNCTIONAL GROUP or ION consisting of a hydrogen ATOM bonded to an oxygen atom (—OH).

I

indicator A substance, such as litmus, which can be used to indicate whether a SOLUTION is ACIDIC or BASIC. Generally, an indicator shows an abrupt color change at a particular PH.

inert gas *See* ARGONON.

infrared radiation Radiation which has a WAVELENGTH

between about 7.5 x 10^{-7}meters and 1 x 10^{-3}meters, that is, the part of the ELECTROMAGNETIC SPECTRUM between the red end of visible light and radio waves. Infrared radiation interacts with different parts of complex MOLECULES in different ways, thus making infrared SPECTROSCOPY a useful analytical tool.

inhibition Prevention of a REACTION from taking place, either wholly or in part. Inhibition is of particular importance in ENZYME-catalyzed reactions, where a substance (the inhibitor) can interact with the enzyme and prevent the substance with which it is meant to react from coming into contact with it.

inorganic chemistry Chemistry of ELEMENTS other than carbon and including COMPOUNDS which contain carbon ATOMS, provided these are not joined to each other. *See also* ORGANIC CHEMISTRY.

ion An ATOM or group of atoms which contains either more or less ELECTRONS associated with it than there are PROTONS. Consequently, all ions carry either positive or negative charges. The two types are known, respectively, as CATIONS and ANIONS. An ion may have more than one unit of charge.

ionic compound A COMPOUND in which the constituent ATOMS exist as IONS and are held together by electrostatic force. *See also* COVALENCY; POLAR MOLECULE.

ionizing radiation Radiation which will interact with matter so that it forms IONS. The short WAVELENGTHS of the ELECTROMAGNETIC SPECTRUM, which include GAMMA and X RAYS, are a form of ionizing radiation, as are some particles, for example, ALPHA PARTICLES. Ionizing radiation is most closely associated with RADIOACTIVE decay.

isomerism The phenomenon whereby the same number and types of ATOMS may join together in different ways, so that more than one distinct chemical COMPOUND can be formed from them. *See also* OPTICAL ACTIVITY; STEREOISOMERISM.

isotope An ATOM of an ELEMENT with a specific number of NEUTRONS. While the number of PROTONS in an atom cannot vary without the atom changing its elemental identity, the number of neutrons can, although only over a small range. Most elements in their natural form are made up from mixtures of two or three different isotopes. Generally, an element has very few stable isotopes; unstable isotopes can break down with the emission of IONIZING RADIATION. *See also* RADIOACTIVITY.

J

joule A unit for measuring quantities of energy. *See also* CALORIE.

K

ketone An ORGANIC chemical substance in which one carbon ATOM has an oxygen atom attached to it by a double BOND and two other carbon atoms attached to it, each by single bonds. *See also* ALDEHYDE.

L

lanthanide One of a group of ELEMENTS which starts with lanthanum (element 57) and concludes with lutetium (element 71). The lanthanides form a separate group in the PERIODIC TABLE in which an inner SHELL of electrons is being filled as the ATOMIC NUMBER increases. *See also* ACTINIDE.

ligand An ATOM or group of atoms capable of donating a pair of ELECTRONS to form a coordinate COVALENT BOND. *See also* COORDINATION COMPOUND; LONE PAIR; TRANSITION ELEMENT.

lipid A triester of glycerol, usually with three molecules of CARBOXYLIC (fatty) acid. Fats and oils from animals and plants are lipids of this type. More complex lipids, in which a phosphoric acid group is linked to the glycerol for example, also occur and play an important role in animal and plant BIOCHEMISTRY. *See also* ESTERIFICATION REACTION; SAPONIFICATION.

lone pair A pair of ELECTRONS in the outer SHELL of an ATOM which are not involved in ordinary COVALENT bonding but are available for donation to a coordinate BOND. *See also* COORDINATION COMPOUND; HYDROGEN BOND.

luminescence The emission of light without the accompaniment of heat, either as the result of some other form of energy being absorbed and re-released (*see* FLUORESCENCE and PHOSPHORESCENCE) or as the result of energy release during a chemical REACTION.

M

macromolecule A MOLECULE made up of many individual ATOMS and thus having a high molecular mass. All POLYMERS are macromolecules.

metal An ELEMENT or mixture of elements which conducts heat and electricity well, and is lustrous, malleable and ductile. In metals, BONDS between ATOMS "resonate," that is, each atom keeps forming and dissolving bonds with neighboring atoms. As a result, the atoms are all closely bound together, but the ELECTRONS are mobile, thus accounting for the electrical conductivity. About 80 per cent of elements are metals. Some elements which lie adjacent to metals in the PERIODIC TABLE show partly metallic and partly nonmetallic characteristics and are called metalloids.

micelle A COLLOIDAL particle which is electrically charged. Micelles are usually made up of aggregates of large MOLECULES.

microwave RADIATION which has a WAVELENGTH between about 1 x 10^{-3}meters and 3 x 10^{-1}meters, that is, covering long INFRARED RADIATION and short radio waves in the ELECTROMAGNETIC SPECTRUM.

mole The quantity of any substance in which the weight in grams is equal to the atomic or molecular mass of the substance. ATOMS, MOLECULES and particles, such as IONS and ELECTRONS, can all be expressed in molar amounts. The concentration of SOLUTIONS is sometimes expressed in terms of molarity (M), in which case it refers to the number of moles of solute per liter of solution. The international symbol for the mole is mol.

molecular biology The branch of science which studies biology at the molecular, or chemical, level. Much of molecular biology can be classed as BIOCHEMISTRY.

molecule Any group of ATOMS linked together by one or more chemical BONDS. A HOMOGENEOUS molecule is one in which the atoms involved are all of the same ELEMENT. Molecules composed of more than one type of atom are called heterogeneous; all chemical COMPOUNDS are composed of such molecules.

monomer A small MOLECULE which, when joined together in large numbers, forms a POLYMER.

N

neutralization The treatment of an ACID with a BASE (or vice versa) to form a SALT in such a way that neither of the original components remains in excess.

neutron One of the three subatomic particles from which

Atoms are made. Like the PROTON, the neutron forms part of the NUCLEUS of an atom. Unlike the proton and the ELECTRON, it carries no electrical charge.

noble gas *See* ARGONON.

nomenclature The system devised for the naming of chemical COMPOUNDS. As the number of known compounds increased, it became necessary to name them according to a system. Systematic nomenclature is now agreed internationally and provides a way of naming MOLECULES so that their structure is unambiguously defined. For many molecules which have been known for a long time, old-fashioned or so-called "trivial" names are still sometimes used.

nuclear fission The breakdown of unstable ISOTOPES, notably of the elements thorium and beyond in the PERIODIC TABLE. In fission, the NUCLEUS of an ATOM breaks into two almost equal sized parts, with the liberation of excess NEUTRONS and large amounts of energy. The unstable isotopes are frequently produced by neutron bombardment; consequently, the release of neutrons by the fission process may start a CHAIN REACTION.

nucleic acid The MACROMOLECULE of which GENETIC material is composed. Nucleic acids are made from NUCLEOTIDE MONOMERS, each comprising one of a small number of nitrogen HETEROCYCLES, a simple SUGAR and a phosphate group. According to whether the sugar is RIBOSE or deoxyribose, the nucleic acids are known as RNA or DNA.

nucleoside *See* NUCLEOTIDE.

nucleotide An ORGANIC chemical COMPOUND made up from one of a small number of nitrogen HETEROCYCLES, a simple SUGAR and a phosphate group. Nucleotides are the MONOMERS from which NUCLEIC ACIDS are made. Some important, small biological molecules, such as ADP and ATP, are nucleotides. The combination of the nitrogen heterocycle and the simple sugar by themselves is called a nucleoside.

nucleus The central part of an ATOM, consisting of PROTONS and NEUTRONS (or, in the case of hydrogen, just a proton). In BIOCHEMISTRY, nucleus may be used to refer to the part of a living cell in which most of the NUCLEIC ACIDS are found.

O

octane number A measure of the performance of automobile fuel. Different HYDROCARBONS burn in slightly different ways. This makes some mixtures more suitable for automobile engines than others. Such mixtures are given an octane number, which compares their burning efficiency against that of the eight-carbon ATOM MOLECULE, *n*-octane, which is arbitrarily given the value 100. Most automobile fuel has octane numbers between 90 and 100.

olefin The former name for an ALKENE.

optical activity The ability of certain MOLECULES to rotate the plane of polarized light. Molecules which are ASYMMETRIC can exist in STEREOISOMERIC forms, the only difference between them being the direction in which they cause the polarized light to rotate. SYNTHESIS of such molecules often produces equal quantities of both ISOMERS, called a racemic mixture. Optical isomerism is of great importance in BIOCHEMISTRY where a biological effect may be produced by one isomeric form, but not the other.

orbital The volume of space in which there is a high probability of finding a particular ELECTRON or pair of electrons. Because the electron has characteristics of both a wave and a particle, its position at any time cannot be determined accurately. It is possible to express its position only in terms of probability. Each electron in an ATOM occupies an atomic orbital. When two electrons link two different NUCLEI in a MOLECULE, they share a volume of space which is called a molecular orbital.

organic chemistry The chemistry of MOLECULES consisting primarily of carbon atoms linked together in chains or rings. At one time it was believed that such substances could be obtained only from matter which was or had been alive, hence the name organic. *See also* INORGANIC CHEMISTRY.

oxidation A REACTION in which the formal electric charge on an ATOM becomes more positive. Although this includes all reactions in which oxygen is added to an ELEMENT or COMPOUND, it is not restricted to reactions involving oxygen. *See also* REDUCTION.

P

paraffin Usually a straight-chain, SATURATED HYDROCARBON or mixture of such hydrocarbons; an ALKANE.

pentose A simple SUGAR containing five carbon ATOMS. RIBOSE, a sugar found in NUCLEIC ACIDS, is a pentose.

peptide A relatively small MOLECULE comprised of a few AMINO ACIDS linked together in the same way as in a PROTEIN. The dividing line between peptides and proteins is not sharply defined.

Periodic Table The basic characterization of modern chemistry, the Periodic Table is a way of arranging ELEMENTS so that those with similar properties form discrete groups.

pH A measure of the ACIDIC or BASIC character of a liquid or SOLUTION, based upon its HYDROGEN ION CONCENTRATION.

pheromone An ORGANIC chemical substance produced by an insect and showing HORMONE-like behavior.

phosphorescence A phenomenon exhibited by various substances in which they absorb light of a particular WAVELENGTH and emit light of a different wavelength. Unlike FLUORESCENCE, which occurs only while the first light source is present, phosphorescence persists after it has been removed. *See also* LUMINESCENCE.

photochemistry The study of chemical REACTIONS which are influenced by light as an energy source.

photosynthesis The production of SUGAR from carbon dioxide and water in plants under the influence of light.

plasma A gas at very high temperature in which the constituent MOLECULES have virtually all dissociated into IONS. This makes the plasma electrically neutral overall but a good conductor of electricity.

plastic Generally, a synthetic ORGANIC POLYMER which can be shaped either during its formation or afterwards by application of heat. Plastics which soften and harden reversibly on heating and cooling are called thermoplastics. The term plastic originally referred to the behavior of any type of material which would deform easily and is sometimes still used in this way.

polar molecule A MOLECULE in which one or more molecular ORBITALS are distorted in shape by the presence of an ELECTRONEGATIVE or ELECTROPOSITIVE atom. As a result, the molecules will align themselves in an electric field as if they had negative and positive poles.

polymer A very large MOLECULE made by linking together many small molecules (MONOMERS), usually of no more than three different types. Polymers derived from a single type of monomer are called homopolymers. *See also* MACROMOLECULE.

polysaccharide A POLYMER formed by linking together

MONOMERS which are simple SUGARS. Polysaccharides occur widely in living organisms, where they are used primarily to store energy. *See also* CELLULOSE; STARCH.

precipitate A solid substance which comes out of a SOLUTION, either as a result of concentrating the solution or of chemical REACTION leading to an insoluble product or products.

protein A POLYMER formed by linking together MONOMERS which are AMINO ACIDS. Proteins play vital roles in living organisms. *See also* ENZYME; PEPTIDE.

proton One of three types of subatomic particle from which ATOMS are made. Together with NEUTRONS, protons form the NUCLEI of atoms. Each proton has a single unit of positive electrical charge which, in a neutral atom, is balanced by the negative charge of an ELECTRON.

Q

quantum number One of four numbers which define the state of an ELECTRON in an ATOM. The numbers indicate the energy level of the electron, the shape of its ORBITAL and its spin. No two electrons in an atom can have all four quantum numbers the same.

R

radiation *See* ELECTROMAGNETIC SPECTRUM; IONIZING RADIATION.

radioactivity The decay of unstable elemental NUCLEI, accompanied by the emission of small particles, such as ALPHA PARTICLES (helium nuclei) or BETA PARTICLES (electrons), and GAMMA RAYS. *See also* HALF-LIFE; ISOTOPE.

rare earth An older name for any of the group of ELEMENTS now called LANTHANIDES.

rare gas *See* ARGONON.

reaction The process by which one or more chemical substances change into one or more different substances.

reagent A chemical substance which is used as a starting material in a chemical REACTION.

reduction A type of REACTION in which the formal electric charge on an ATOM becomes more negative. *See also* OXIDATION.

ribose A simple PENTOSE SUGAR which forms part of the NUCLEOTIDE MONOMERS from which ribonucleic acid (RNA) is made.

ribosome A small particle found in living cells which is the site of PROTEIN synthesis.

RNA Ribonucleic acid, a POLYMERIC material found in most living organisms which transfers the genetic code (*see* GENE) embedded in DNA to the RIBOSOMES. In some VIRUSES, the RNA is the genetic material rather than DNA. *See also* NUCLEIC ACID.

S

salt A chemical substance which exists as an IONIC solid. Salts are formed when an ACID and a BASE NEUTRALIZE each other. Most salts are composed of METAL CATIONS and ANIONS of ELECTRONEGATIVE elements, such as chlorine. Common or table salt is sodium chloride.

saponification The breakdown of an ESTER into an ALCOHOL and the SALT of a CARBOXYLIC ACID in the presence of a BASE. The name derives from the original method for producing soap, which involves the reaction between TRIGLYCERIDES (in naturally-occurring fats and oils) and a metallic base.

saturation The involvement of all bonding ELECTRONS of an

ATOM, usually carbon, in single BONDS with other atoms. Also, the state of a SOLUTION which can dissolve no more SOLUTE.

shell A layer of ELECTRONS surrounding an ATOM at approximately the same distance from the NUCLEUS. Electrons in different shells are at different average distances from the nucleus and it is usually only those electrons in the outermost shell which participate in chemical REACTIONS. The first QUANTUM NUMBER indicates the shell in which an electron occurs.

solubility A measure of the extent to which a particular substance will dissolve in a particular liquid. Solubility often changes with temperature, so that a SATURATED SOLUTION at high temperature will often PRECIPITATE material as it cools. This can often be used as a method of purifying substances.

solute A substance, usually solid, which is dissolved in a liquid (the SOLVENT) to form a SOLUTION. *See also* SOLUBILITY.

solution A HOMOGENEOUS mixture of two or more different substances, usually a solid in a liquid.

solvent A liquid in which a substance (the SOLUTE) is dissolved to form a SOLUTION. *See also* SOLUBILITY.

specific gravity A measure of the density of a material relative to that of pure water at 40°F (4°C).

spectroscopy The study of the ABSORPTION or emission of ELECTROMAGNETIC RADIATION by substances. Spectroscopy provides chemists with a range of techniques for identifying compounds according to their specific spectral characteristics. It is divided into several different types, such as INFRARED and ULTRAVIOLET, according to the WAVELENGTHS of the ELECTROMAGNETIC SPECTRUM which are used.

spectrum *See* ELECTROMAGNETIC SPECTRUM.

spin One of the quantum properties of subatomic particles, related to the way in which they spin on their axes (clockwise or anticlockwise). The spin of an ELECTRON is its fourth QUANTUM NUMBER. Spin is of particular importance in some types of SPECTROSCOPY.

starch A CARBOHYDRATE POLYMER found widely in plants, which use it as a means of storing energy.

stereoisomer One form of a chemical COMPOUND which can exist in distinct, OPTICALLY ACTIVE forms because one of the ATOMS has different substituents attached to it in such a way that a nonsuperimposable mirror-image structure is also possible.

steroid One of a class of biologically active COMPOUNDS characterized by a skeleton made up of four fused carbon rings. Steroids are important as VITAMINS and HORMONES.

sublimation The process in which a solid turns to a vapor without passing through an intermediate liquid state. Whether this can happen or not depends on the substance involved and the temperature and pressure to which it is subjected. Few substances sublime at ordinary atmospheric pressure and moderate temperatures. Elemental iodine is one that does.

sugar A small CARBOHYDRATE MOLECULE; often used to refer to the compound sucrose.

superconductivity A phenomenon observed primarily at extremely low temperatures in which the electrical CONDUCTIVITY of a substance increases dramatically.

surfactant A MOLECULE which affects the behavior of liquids at interfaces. Commonly, surfactants are molecules which have a HYDROPHILIC and a HYDROPHOBIC part. As a result, they can interact with both water and water-immiscible liquids, forming EMULSIONS.

synthesis The process by which complex MOLECULES are made from simpler molecules.

T

tautomerism The interconversion of two ISOMERIC forms of a MOLECULE. Some substances can exist in different isomeric forms which interchange readily so that an EQUILIBRIUM concentration of each exists.

titration The determination of the strength or nature of a SOLUTION by the addition of a solution containing a specific CONCENTRATION of a known substance which will react in some way with the SOLUTE in the unknown solution. The most common use of titration is to determine the strength of an ACID solution by titration with a BASE, or vice versa. In such cases, an INDICATOR is used to show when all the unknown material has been NEUTRALIZED.

transition elements Elements which have an incomplete ELECTRON SHELL and one or more electrons in the next higher shell. Such elements, which occupy the middle of the PERIODIC TABLE, readily form coordinate BONDS with atoms which have LONE PAIRS of electrons. *See also* LIGAND.

transition state The intermediate state in a chemical REACTION between reactants and products. The transition state is at a higher energy than either reactants or products and its formation represents an energy barrier which has to be overcome if reaction is to occur. *See also* ACTIVATION ENERGY.

triglyceride A naturally-occurring fat or oil which is an ESTER formed from three molecules of CARBOXYLIC ACID and a molecule of glycerol. *See also* LIPID.

U

ultraviolet radiation ELECTROMAGNETIC radiation which has a WAVELENGTH between about 1×10^{-9} meters and 1×10^{-7} meters. The long end of the ultraviolet waveband abuts the short end of the visible spectrum, which is violet. The ultraviolet wavelengths are important in SPECTROSCOPY and also strongly influence some chemical REACTIONS.

unsaturated compound An ORGANIC chemical COMPOUND which contains at least two carbon ATOMS joined by a multiple BOND. *See also* UNSATURATION.

unsaturation The state of an ATOM in which it is bound to fewer other atoms than is theoretically possible, assuming a single (two-electron) BOND between each pair of atoms. Most frequently, it denotes the linking of carbon atoms by double or triple bonds which can be broken so as to leave a single bond still holding the two atoms together.

V

vaccine A preparation which, when introduced into the bloodstream, causes the production of ANTIBODIES against a particular infectious organism. Vaccines are produced from weakened strains of the organism or dead cells.

valence The combining power of an ATOM. The valence ELECTRONS are usually those in the outermost SHELL, which can participate in BOND formation with other atoms. The valence of an atom is often equal to the number of electrons it needs to lose or gain — whichever is the less — in order to achieve a completed shell. However, many elements, particularly the TRANSITION ELEMENTS, show variable valency.

virus A particle which is capable of taking over part of the mechanism of a living cell in order to reproduce itself. Many viruses consist solely of RNA and PROTEIN and are responsible for causing various diseases.

vitamin A substance which is needed by the body in small quantities if it is to function efficiently, but which the body cannot usually synthesize for itself. Deficiency of a particular vitamin induces specific disease symptoms. Most vitamins are complex ORGANIC compounds which act as COENZYMES.

W

wavelength A property of ELECTROMAGNETIC radiation. Such radiation has wavelike characteristics and the wavelength is the distance between adjacent similar points in the wave function. Wavelengths in the electromagnetic spectrum vary between about 10 trillionths of a meter (10^{-17} m) for the shortest cosmic rays to 10,000 meters for the longest radio waves. In customary units, this is equivalent to a range of about 400 trillionths of an inch up to about 6 miles.

X

X rays ELECTROMAGNETIC radiation which has a WAVELENGTH between that of GAMMA RAYS and ULTRAVIOLET RADIATION.

INDEX

In this index, italic numerals (e.g. *118*) refer to illustrations or their captions. The symbol * indicates a topic that has an entry in the Glossary (pp. 144—151).

A

CREDITS

Air Products Ltd 57; Airship Industries 56; Heather Angel 35, 39, 67, 84, 90, 113; St Bartholomew's Hospital 31; De Beers 36; Biophotos Associates 109, 120; B. P. Chemicals 66; Paul Brierley 20, 35, 43, 76, 78, 109, 110, 137, 140, 141, 143; Brinsley Burridge/Nature Photographers Ltd 96, 123; British Aerospace Dynamics Group 91; British Steel Corporation 63; British Sulphur Corporation 53; Dr J. Burgess/Science Photo Library 47; Camerapix Hutchison Library 86; W. Canning Materials Ltd 64; M. Chinery/Natural Science Photos 78; Cleveland Potash/Agricultural Publicity 16; Bruce Coleman Ltd 92; R. Constable/Camerapix Hutchison Library 63; Courtaulds Ltd 21; The Design Council 98; The Distillers Co 80; Division of Computer Research and Technology/National Institute of Health/Science Photo Library 127; Martin Dohrn/Science Photo Library 14; Dunlop Sports Company Ltd 37; Dr Harold Edgerton/Science Photo Library 19; Sarah Errington/Camerapix Hutchison Library 97; Barry Finch/Courtaulds 99; Vaughan Fleming/Science Photo Library 95; Institute of Geological Sciences 22, 39, 55; Geoscience Features 58, 117; Gold Information Council 29; Gower Medical Publishing 80; Goycolea/Camerapix Hutchison 39; Robert Harding Associates 109; J. Heinz Ltd 38; J. Hobday/Natural Science Photos 9; Holt Studios Photographic Library 93, 107, 143; Hong Kong Tourist Association 94; Johnson Matthey 44; David Jones 131; Kodak 28; J. Laing Ltd 33; Jane Mackinnon/Seaphot 30; G. Montalverne/Natural Science Photos 96; Sidney Moulds/Science Photo Library 3; Dr Gopal Murti/Science Photo Library 125; NASA/Science Photo Library 50, 61; U.S. Navy/Science Photo Library 24; Peter Newarks Western Americana 25; Martyn Page 8, 28, 36, 45, 69, 77, 97; Perkin-Elmer Ltd 130, 132, 135, 138, 139, 140; Photosource 9, 25, 40, 42, 48, 68, 71, 77, 87, 88, 101, 107; Platinum Information/Courtesy Goldsmith's Hall 45; J. von Puttkamer/Camerapix Hutchison Library 33; Redifussion 58; The Royal Institution 18; RTZ 32; The Sausage Bureau 27; Science Photo Library 62; Dr G. P. Schatten/Science Photo Library 113; Science Source/Science Photo Library 124; Shell 47, 103; D. Simpson/Camerapix Hutchison Library 49; Tony Stone Associates 10, 11, 15, 19, 20, 27, 30, 34, 43, 47, 51, 55, 57, 65, 67, 72, 73, 74, 81, 83, 85, 87, 89, 102, 105, 111, 112, 116, 123, 124, 127, 128; Thames Water Authority 142; Steve Thompson 31; Unilever Educational Publications 73; UKAEA 61; Vautier De Nauxe 75; John Walsh/Science Photo Library 29; P. H. and S. L. Ward/Natural Science Photos 122, 84; M. Watson/Biofotos 69; J. Welchman 78, 79, 83; Werner Forman Archive/National Museum of Copenhagen 63; Peter Westbury/Meteorological Office 127.

Cover photograph courtesy of ZEFA, Düsseldorf, West Germany